GRADE

everyday Mathematics

Intervention Activities

Table of Contents

Using Everyday Mathematics Intervention Activities

Current research indicates that literacy activities that engage students in familiar, real-world math situations are essential for math skill development. The Everyday Mathematics Intervention Activities series offers activities that are carefully crafted to help students grow in language and literacy while acquiring core grade-level math content.

Effective mathematics activities provide students with opportunities to:

- Strengthen the language and literacy skills needed to develop math proficiency
- Relate math concepts to real-life situations
- Develop math computation and application skills

Although some students master these skills easily during regular classroom instruction, many others need additional re-teaching opportunities to master these essential skills. The Everyday Mathematics Intervention Activities series provides easy-to-use, five-day intervention units for Grades K–5. These units are structured around a research-based Model-Guide-Practice-Apply approach. You can use these activities in a variety of intervention models, including Response to Intervention (RTI).

Getting Started

In just five simple steps, Everyday Mathematics Intervention Activities provides everything you need to identify students' needs and to provide targeted intervention.

1. PRE-ASSESS to identify students' mathematics needs. Use the pre-assessment to identify the skills your students need to master.

online

Pretest: Polygons

2. MODEL the skill. Every five-day unit targets a specific mathematics area. On Day 1, use the teacher prompts and reproducible activity page to introduce and model the skill.

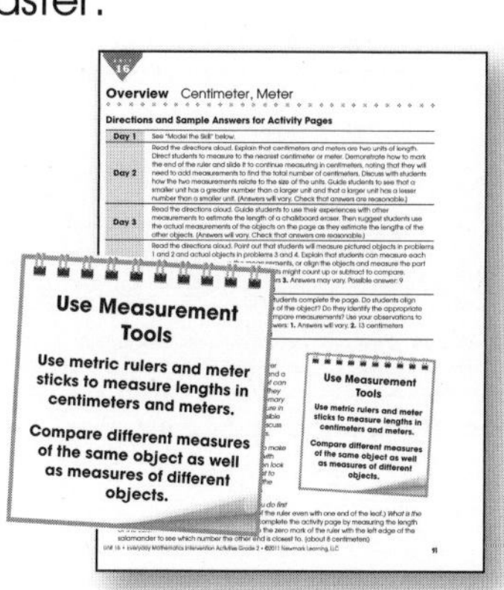
Overview Centimeter, Meter

Directions and Sample Answers for Activity Pages

Use Measurement Tools

Use metric rulers and meter sticks to measure lengths in centimeters and meters.

Compare different measures of the same object as well as measures of different objects.

Day 1

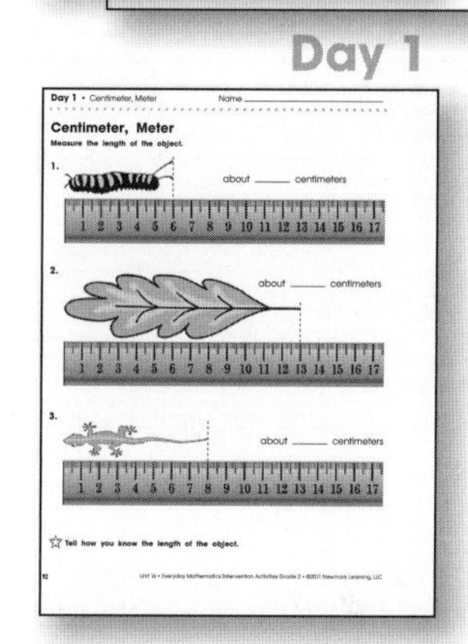
Centimeter, Meter

Measure the length of the object.

3. GUIDE, PRACTICE, and APPLY. Use the reproducible practice activities for Days 2, 3, and 4 to build students' understanding and skill proficiency.

Day 2 Day 3 Day 4

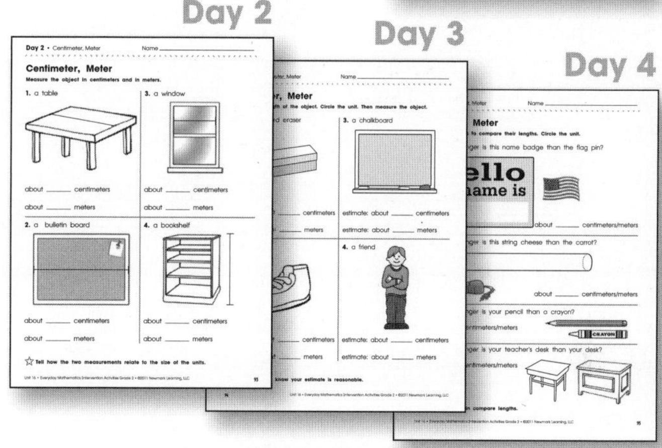
Centimeter, Meter

Measure the object in centimeters and in meters.

4. MONITOR progress. Administer the Day 5 reproducible assessment to monitor each student's progress and to make instructional decisions.

Day 5

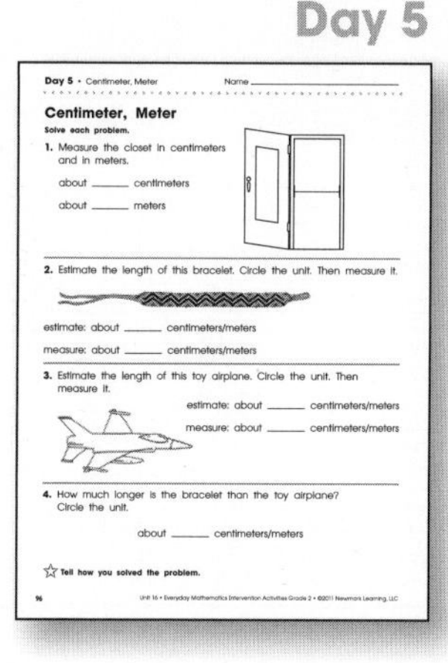
Centimeter, Meter

Solve each problem.

5. POST-ASSESS to document student progress. Use the post-assessment to measure students' progress as a result of your interventions.

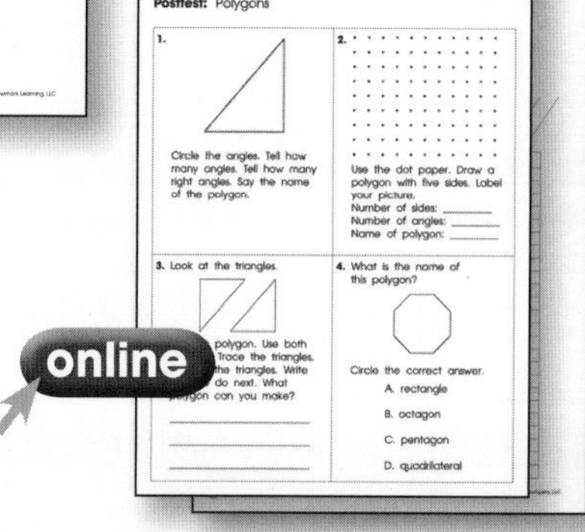
Posttest: Polygons

online

Standards-Based Mathematics Awareness Skills in Everyday Intervention Activities

The mathematics strategies found in the Everyday Intervention Activities series are introduced developmentally and spiral from one grade to the next. The chart below shows the types of skill areas addressed at each grade level in this series.

Everyday Mathematics Intervention Activities Series Words	K	1	2	3	4	5
Counting & Cardinality	✔					
Number & Operations	✔	✔	✔	✔	✔	✔
Algebraic Thinking	✔	✔	✔	✔	✔	✔
Fractions				✔	✔	✔
Measurement & Data	✔	✔	✔	✔	✔	✔
Geometry	✔	✔	✔	✔	✔	✔

Using Everyday Intervention for RTI

According to the National Center on Response to Intervention, RTI "integrates assessment and intervention within a multi-level prevention system to maximize student achievement and to reduce behavior problems." This model of instruction and assessment allows schools to identify at-risk students, monitor their progress, provide research-proven interventions, and "adjust the intensity and nature of those interventions depending on a student's responsiveness."

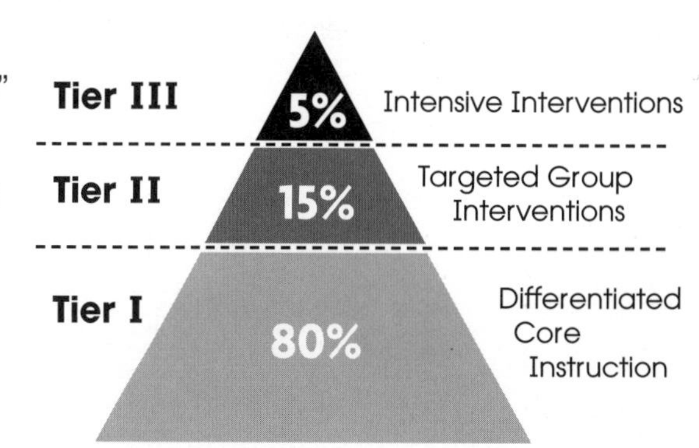

RTI models vary from district to district, but the most prevalent model is a three-tiered approach to instruction and assessment.

The Three Tiers of RTI	Using Everyday Intervention Activities
Tier I: Differentiated Core Instruction • Designed for all students • Preventive, proactive, standards-aligned instruction • Whole- and small-group differentiated instruction • Ninety-minute, daily core reading instruction in the five essential skill areas: Mathematics, phonemic awareness, comprehension, vocabulary, fluency	• Use whole-group math mini-lessons to introduce and guide practice with math strategies that all students need to learn. • Use any or all of the units in the order that supports your core instructional program.
Tier II: Targeted Group Interventions • For at-risk students • Provide thirty minutes of daily instruction beyond the ninety-minute Tier I core instruction • Instruction is conducted in small groups of three to five students with similar needs	• Select units based on your students' areas of need (the pre-assessment can help you identify these). • Use the units as week-long, small-group mini-lessons.
Tier III: Intensive Interventions • For high-risk students experiencing considerable difficulty • Provide up to sixty minutes of additional intensive intervention each day in addition to the ninety-minute Tier I core instruction • More intense and explicit instruction • Instruction conducted individually or with smaller groups of one to three students with similar needs	• Select units based on your students' areas of need. • Use the units as one component of an intensive math intervention program.

Overview Addition and Subtraction Fact Families

Directions and Sample Answers for Activity Pages

Day 1	See "Model the Skill" below.
Day 2	Read the directions aloud. Remind students that a related fact uses the same three numbers. If students have difficulty determining the related addition fact, have them build the cube train and turn it around to see that the order of the parts changes, but the total does not change. Have students write one of the related subtraction facts. (Answers: **1.** 4 + 6 = 10, 10 – 6 = 4 or 10 – 4 = 6; **2.** 3 + 4 = 7, 7 – 4 = 3 or 7 – 3 = 4; **3.** 2 + 7 = 9, 9 – 2 = 7 or 9 – 7 = 2; **4.** 3 + 5 = 8, 8 – 5 = 3 or 8 – 3 = 5)
Day 3	Read the directions aloud. Point out that students can use the one given fact to determine a related addition fact and two related subtraction facts. Guide students to switch the addends and then turn around the addition facts to help determine the subtraction facts. (Answers: **1.** 8 + 7 = 15, 15 – 8 = 7, 15 – 7 = 8; **2.** 4 + 9 = 13, 13 – 9 = 4, 13 – 4 = 9; **3.** 6 + 8 = 14, 14 – 6 = 8, 14 – 8 = 6; **4.** 7 + 9 = 16, 16 – 7 = 9, 16 – 9 = 7)
Day 4	Read the directions aloud. Point out that all three numbers given should be in each fact. Remind students that an addition fact ends with the greatest number while a subtraction fact begins with the greatest number. (Answers: **1.** 9 + 8 = 17, 8 + 9 = 17, 17 – 8 = 9, 17 – 9 = 8; **2.** 5 + 7 = 12, 7 + 5 = 12, 12 – 7 = 5, 12 – 5 = 7; **3.** 6 + 9 = 15, 9 + 6 = 15, 15 – 9 = 6, 15 – 6 = 9; **4.** 7 + 6 = 13, 6 + 7 = 13, 13 – 6 = 7, 13 – 7 = 6)
Day 5	Read the directions aloud. Observe as students complete the page. Do students reverse the order of the numbers in the addition fact when they write the related subtraction fact? Do they use all three numbers in each fact of the fact family? Use your observations to plan further instruction. (Answers: **1.** 12 – 6 = 6; **2.** 3 + 7 = 10, 10 – 7 = 3 or 10 – 3 = 7; **3.** 5 + 4 = 9, 4 + 5 = 9, 9 – 4 = 5, 9 – 5 = 4; **4.** 4 + 8 = 12, 8 + 4 = 12, 12 – 8 = 4, 12 – 4 = 8)

Model the Skill

- Hand out the Day 1 activity page and connecting cubes. **Ask:** *How many cubes are darker in the first problem?* (5) *How many cubes are lighter?* (4) Have students make a cube train with 5 cubes of one color and 4 cubes of another color. **Say:** *The cubes show the addition fact 5 + 4 = 9. Break the cube train into two colors. A related subtraction fact uses the same numbers as the addition fact. What related subtraction fact did you show?* (9 – 4 = 5 or 9 – 5 = 4) Have students record one of the facts.

- **Say:** *Make a cube train like the one shown in problem 2. Compare your cube train to the addition problem shown. What is the total number of cubes?* (10) *Where do you write the total in a subtraction fact?* (Possible answer: It is written first.) Observe as students break their cube train into two colors and record one of the related subtraction facts. (10 – 8 = 2 or 10 – 2 = 8)

- Help students complete the activity page using cube trains and addition facts to identify related subtraction facts. Remind students to start with a total, subtract one color, and then write the difference. (Answers: **3.** 8 – 1 = 7 or 8 – 7 = 1; **4.** 9 – 3 = 6 or 9 – 6 = 3)

Addition and Subtraction Fact Families

Write a related subtraction fact.

1.

5 + 4 = 9 ______ – ______ = ______

2.

8 + 2 = 10 ______ – ______ = ______

3.

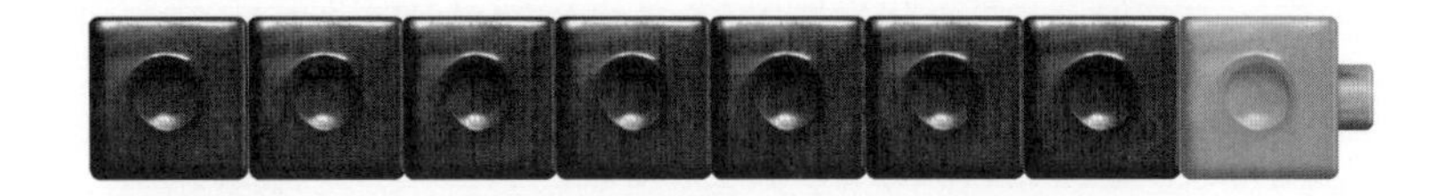

7 + 1 = 8 ______ – ______ = ______

4.

6 + 3 = 9 ______ – ______ = ______

☆ **Tell how to find a related subtraction fact when given an addition fact.**

 Name ____________________

Addition and Subtraction Fact Families

Write the related addition fact. Then write a related subtraction fact.

1

6 + 4 = 10 ______ + ______ = ______

______ − ______ = ______

2

4 + 3 = 7 ______ + ______ = ______

______ − ______ = ______

3

7 + 2 = 9 ______ + ______ = ______

______ − ______ = ______

4

5 + 3 = 8 ______ + ______ = ______

______ − ______ = ______

☆ **Tell how to find a related addition fact.**

Addition and Subtraction Fact Families

Write the related addition and subtraction facts.

1

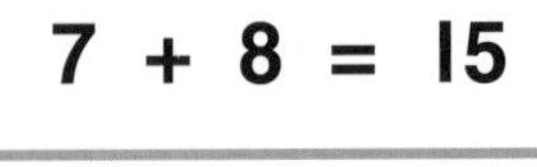

____ + ____ = ____

____ − ____ = ____

____ − ____ = ____

2

9 + 4 = 13

____ + ____ = ____

____ − ____ = ____

____ − ____ = ____

3

8 + 6 = 14

____ + ____ = ____

____ − ____ = ____

____ − ____ = ____

9 + 7 = 16

____ + ____ = ____

____ − ____ = ____

____ − ____ = ____

☆ **Tell how you can find the related addition and subtraction facts.**

Addition and Subtraction Fact Families

Use the numbers shown to write the facts in the fact family.

1

17
9 8

____ + ____ = ____ ____ + ____ = ____

____ – ____ = ____ ____ – ____ = ____

2

12
5 7

____ + ____ = ____ ____ + ____ = ____

____ – ____ = ____ ____ – ____ = ____

3

15
6 9

____ + ____ = ____ ____ + ____ = ____

____ – ____ = ____ ____ – ____ = ____

4

13
7 6

____ + ____ = ____ ____ + ____ = ____

____ – ____ = ____ ____ – ____ = ____

☆ **Tell how you know the facts in a fact family.**

Assessment

Write a related subtraction fact.

6 + 6 = 12 ________ – ________ = ________

Write the related addition fact. Then write a related subtraction fact.

7 + 3 = 10 ________ + ________ = ________

________ – ________ = ________

Use the numbers shown to write the facts in the fact family.

3

9 5 4	____ + ____ = ____ ____ + ____ = ____	____ – ____ = ____ ____ – ____ = ____

4

12 4 8	____ + ____ = ____ ____ + ____ = ____	____ – ____ = ____ ____ – ____ = ____

☆ **Tell how you solved the problem.**

Overview Write a Number Sentence

Directions and Sample Answers for Activity Pages

Day 1	See "Model the Skill" below.
Day 2	Read the directions aloud. Encourage students to draw simple pictures, such as lines or dots, to show the problems. Guide them to determine if they will add or subtract to solve each problem before they make their drawings. Point out that students should draw and compare in problem 1. (Answers: **1.** 12 – 9 = 3; **2.** 8 + 7 = 15; **3.** 17 – 8 = 9; **4.** 9 + 4 = 13)
Day 3	Read the directions aloud. Explain that often the unknown is the sum or the difference, but it could be in any position in a number sentence. Show students how the unknown in problem 1 could either the difference or a missing addend. Tell students that the first number sentence in each problem should include a square for the unknown and the second number sentence should be complete. (Possible answers: **1.** 15 – 6 = 9; **2.** 7 + 4 = 11; **3.** 8 + 6 = 14; **4.** 8 – 7 = 1)
Day 4	Read the directions aloud. Point out that the problems are multi-step and will require students to write two number sentences to solve. Encourage students to read the entire problem and think about what operations they will use before they write number sentences. (Answers: **1.** 4 + 2 = 6, 6 – 3 = 3; **2.** 8 + 3 = 11, 11 – 4 = 7; **3.** 7 – 3 = 4, 4 – 2 = 2; **4.** 10 – 2 = 8, 8 – 4 = 4)
Day 5	Read the directions aloud. Observe as students complete the page. Do students choose the correct operation? Do they perform both steps in order in a multi-step problem? Use your observations to plan further instruction and review. (Answers: **1.** 6 + 3 = 9; **2.** 11 – 6 = 5; **3.** 7 – 4 = 3; **4.** 4 + 6 = 10, 10 – 2 = 8)

Model the Skill

- Hand out the Day 1 activity page and 10 counters of one color and 10 counters of another color to each student. Invite a student to read aloud problem 1. **Say:** *You can show a math story problem with objects or pictures. Today we will use counters. How many counters should we show to represent the red apples?* (5) *How many counters should we show to represent the green apples?* (3) Have students use different colors for different color apples.
- **Ask:** *What is the question asking us to find out?* (how many apples there are altogether) *Do you add the groups or compare the groups to find out?* (add) Observe as students put the groups together to find the total number of apples. **Ask:** *What number sentence shows this problem?* (5 + 3 = 8) Observe as students complete the number sentence for problem 1. Guide them to write the symbols in the circles and the numbers on the lines.
- Invite a student to read aloud problem 2. **Ask:** *Will you add or subtract to find the answer to this problem?* (subtract) *How do you know?* (Possible answer: The question asks how many oranges are left.) *How many counters will you use to show Mary's oranges?* (7) *What will you do with those counters?* (Take away 6.) Have students solve the problem with counters and write the matching number sentence.
- Help students complete the page using counters to model each problem. Observe as they decide whether to add or subtract and complete each number sentence. (Answers: **3.** 9 – 5 = 4; **4.** 6 + 4 = 10)

Use Manipulatives

Use counters to model the word problems.

Decide if counters should be added, subtracted, or compared.

Record the number sentence that matches the model.

Name ______________________________

Write a Number Sentence

Solve with counters. Complete each number sentence.

1. There are 5 red apples and 3 green apples.

   How many apples are there all together?

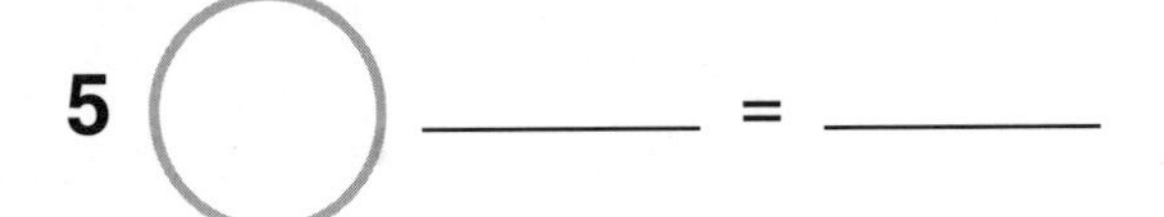

2. Mary has 7 oranges.

   She uses 6 to make juice.

   How many oranges does she have left?

7 ◯ ______ = ______

3. Jordan has a bowl of 9 strawberries.

   He puts them into 2 groups. There are 5 strawberries in one group.

   How many strawberries are in the other group?

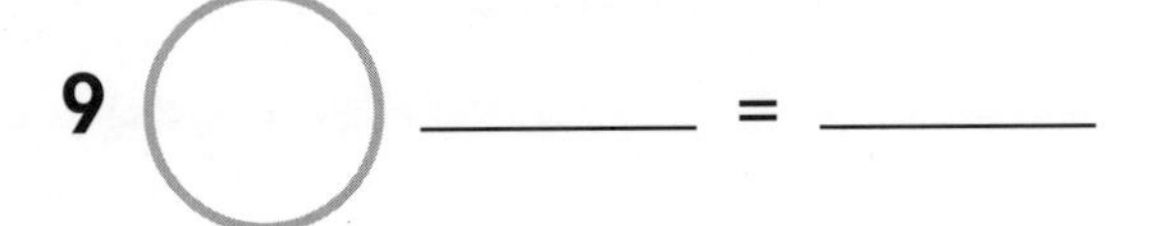

4. Jan has 6 peaches.

   Dana gives her 4 more peaches.

   How many peaches does Jan have now?

6 ◯ ______ = ______

☆ **Tell how you know whether to add or subtract.**

Name ____________________

Write a Number Sentence

Draw a picture to solve. Write the number sentence.

1. There are 12 silver sequins and 9 gold sequins. How many fewer gold sequins are there than silver sequins?

   ____ ◯ ____ ◯ ____

2. Wendy has 8 yellow beads and 7 green beads. How many beads in all does Wendy have?

   ____ ◯ ____ ◯ ____

3. Carl bought 17 buttons. He used 8 on a shirt. How many buttons does he have left?

   ____ ◯ ____ ◯ ____

4. Dawn made a wreath with 9 flowers. Then she put 4 more flowers on the wreath. How many flowers are on the wreath now?

   ____ ◯ ____ ◯ ____

☆ **Tell how you know what number sentence to write.**

 Name ____________________

Write a Number Sentence

Use a ☐ for the unknown. Then solve.

1. Hannah has 15 shirts.

   There are 6 in her suitcase. The rest are in her drawer.

   How many shirts are in Hannah's drawer?

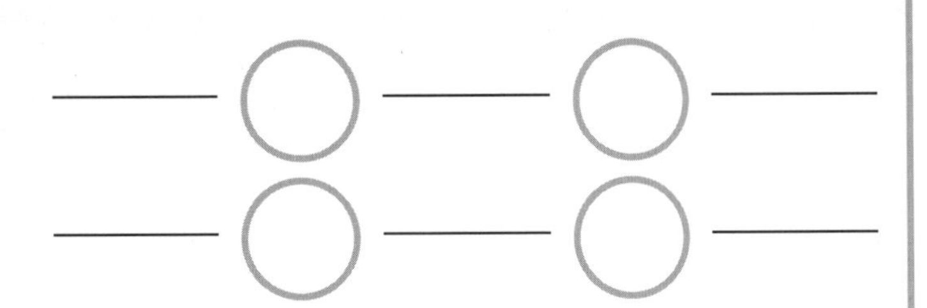

2. Mr. West has 7 ties with stripes and 4 ties with dots.

   How many ties does he have in all?

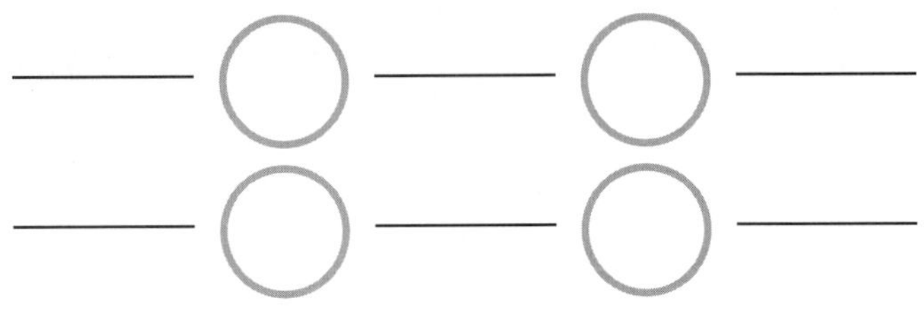

3. Becky has 8 pairs of socks.

   Her mother buys her 6 more pairs of socks.

   How many pairs of socks does Becky have now?

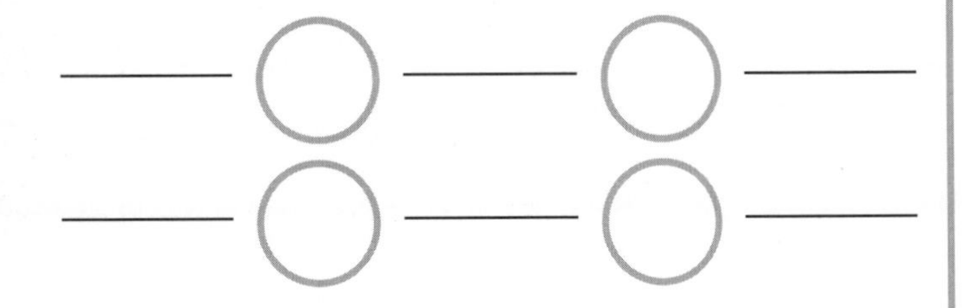

4. Tommy has 8 baseball caps.

   Doug has 7 baseball caps.

   How many more caps does Tommy have than Doug?

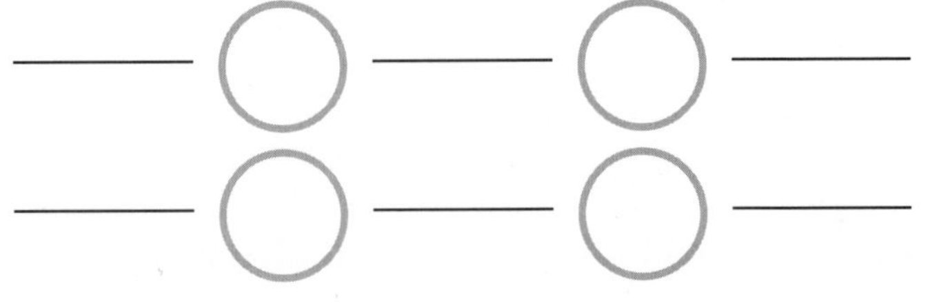

☆ **Tell how you know which amount is unknown.**

 Name ______________________

Write a Number Sentence

Write the number sentences. Solve.

1. Mrs. Walsh bought 4 blueberry yogurts and 2 strawberry yogurts.
Her 3 children each ate a yogurt for snack.

How many yogurts are left?

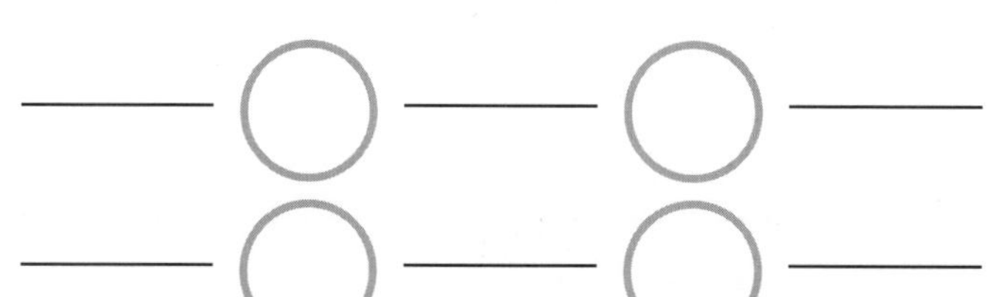

2. Lisa put 8 slices of American cheese and 3 slices of Swiss cheese on a platter.
Her sons took 4 slices of cheese to make sandwiches.

How many slices of cheese are left on the platter?

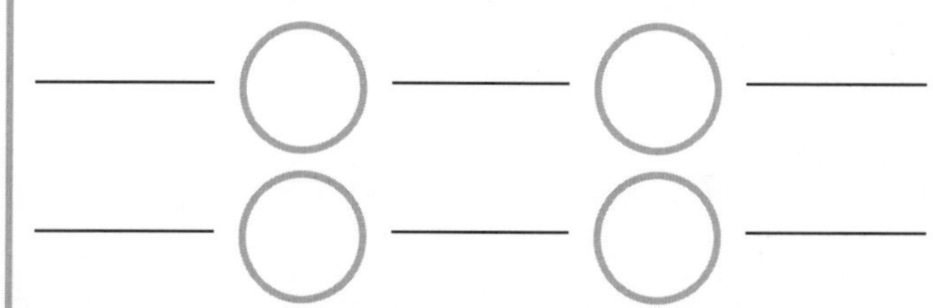

3. There were 7 crackers on a plate.
Sam ate 3 crackers.
Then Rachel ate 2 crackers.

How many crackers are still on the plate?

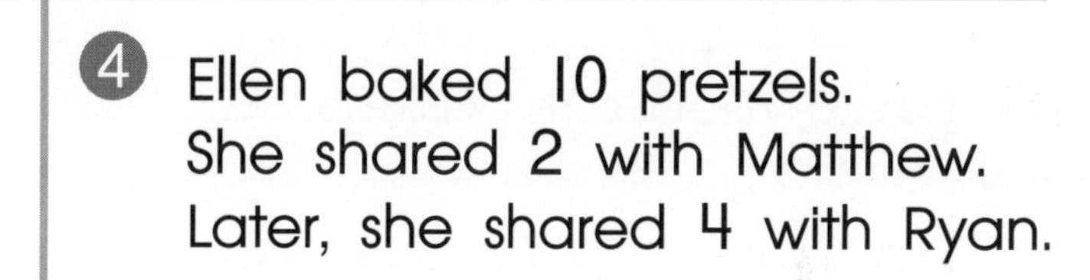

4. Ellen baked 10 pretzels.
She shared 2 with Matthew.
Later, she shared 4 with Ryan.

How many pretzels does Ellen have now?

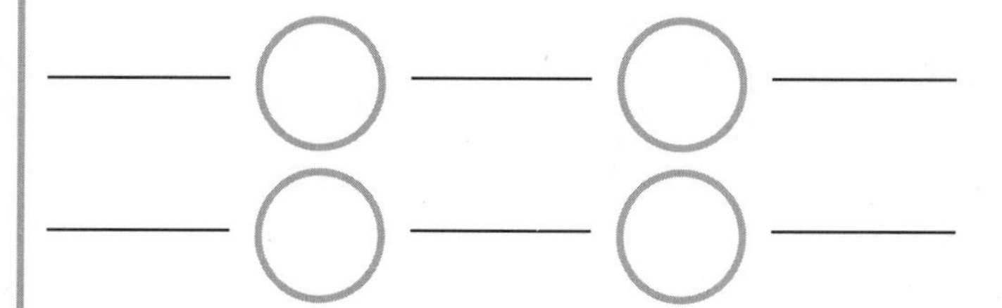

☆ **Tell how you solved the problem.**

Assessment

Write a number sentence to solve each problem.

❶ The Rockets made 6 goals in the first half of the soccer game.

They made 3 goals in the second half of the game.

How many goals did they make in all?

6 ○ ______ = ______

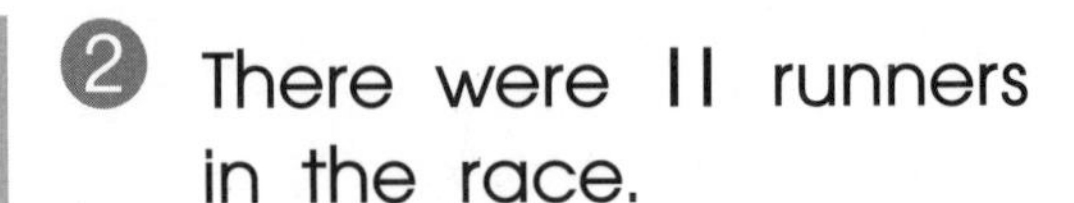

❷ There were 11 runners in the race.

6 runners wore red shirts. The other runners wore blue shirts.

How many runners wore blue shirts?

______ ○ ______ ○ ______

❸ Nick made 7 baskets.

Eddie made 4 baskets.

How many more baskets did Nick make?

______ ○ ______ ○ ______

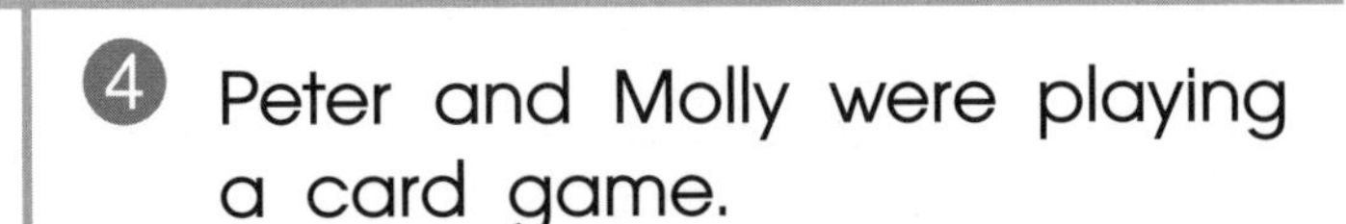

❹ Peter and Molly were playing a card game.

Peter scored 4 points in the first round. He then scored 6 points in the second round.

Then, in the third round, he lost 2 points.

What was Peter's score after the third round?

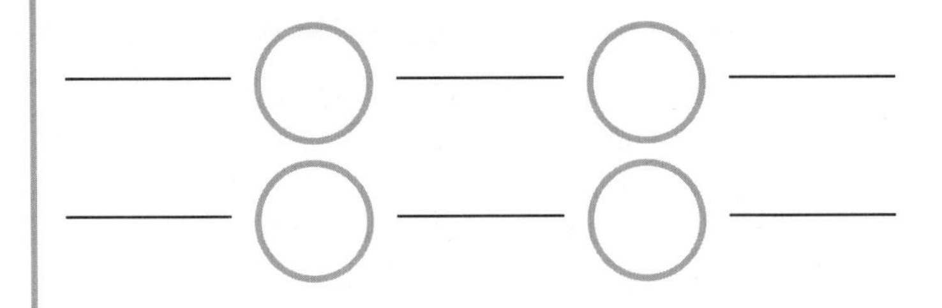

______ ○ ______ ○ ______

______ ○ ______ ○ ______

☆ **Tell how you solved the problem.**

Overview Odd and Even

Directions and Sample Answers for Activity Pages

Day 1	See "Model the Skill" below.
Day 2	Read the directions aloud. Remind students that a pair is two objects. Allow students to circle pairs in any way they like. Then guide them to see that the counters are aligned vertically so that they can easily see in which problems all the counters pair to show an even number. (Answers: **1.** even; **2.** odd; **3.** odd; **4.** even)
Day 3	Read the directions aloud. Point out that the objects on this page are not pairs visually so students will have to circle pairs. Encourage students to start at the left of each problem and circle pairs. Remind students that they should write **odd** when there is an unpaired object and **even** when every object has a partner. (Answers: **1.** even; **2.** even; **3.** odd; **4.** odd)
Day 4	Read the directions aloud. Point out that the cube towers are arranged so that the cubes in one tower partner with the cubes in the other tower. Guide students to start at the bottom of the towers and draw horizontal lines to connect cubes in each tower and, in essence, form a ladder. Students should note that only even sums are shown. (Answers: **1.** 2 + 2 = 4; **2.** 3 + 3 = 6; **3.** 7 + 7 = 14; **4.** 6 + 6 = 12)
Day 5	Read the directions aloud. Have students complete the page. Do students circle pairs starting on the left? Do they notice the pairings created by the arrangement of items in rows and columns? Can they identify that the sum of two equal addends is even? Use your observations to plan further review. (Answers: **1.** odd; **2.** odd; **3.** even; **4.** 8 + 8 = 16)

Model the Skill

- Hand out the Day 1 activity page and 10 connecting cubes to each student. **Ask:** *How many cubes are in the train in problem 1?* (5) Have students each make a cube train using 5 cubes. **Ask:** *How many are in a pair?* (2) *To find out if your cube train has an odd or even number, you can pair the cubes.* Guide them to break the train into pairs of cubes. To reinforce the cube pairs, have students draw a vertical line between every two cubes. **Ask:** *Does every cube have a partner?* (no) **Say:** *If there is a cube without a partner, it means that the number of cubes is odd.*
- **Say:** *Make a cube train to match the number of cubes in problem 2. Then break the train into cube pairs.* Help students create their train and break it into 3 cube pairs. **Ask:** *Does every cube have a partner?* (yes) *If every cube has a partner, it means that the number of cubes is even.* Observe as students circle the word **even**.

- Help students complete the activity page by making each cube train, breaking the train into pairs of cubes, and deciding whether the number of cubes is even or odd. Remind students that a cube without a partner means that the number is odd. (Answers: **3.** odd; **4.** even)

Name ______________________________

Odd and Even

Break the train into cube pairs. Circle whether the number of cubes is odd or even.

1

odd even

2

odd even

3

odd even

4

odd even

☆ **Tell how you know an amount is even.**

 Name ______________________

Odd and Even

Circle pairs. Then circle whether the number is odd or even.

1

odd even

2

odd even

3

odd even

4

odd even

☆ **Tell how you know when a number is odd.**

 Name ________________

Odd and Even

Circle pairs. Then write whether the amount is odd or even.

____

2

____

3

____

4

____

☆ **Tell how you know whether to write *odd* or *even*.**

 Name ____________________

Odd and Even

Draw lines to connect pairs. Write the number in each tower. Then add.

1

_______ + _______ = _______

2

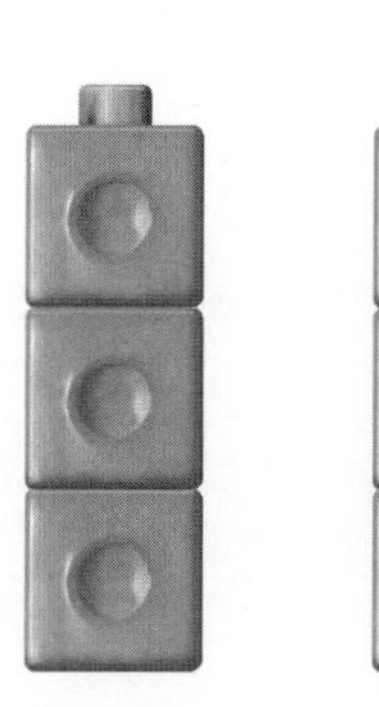

_______ + _______ = _______

3

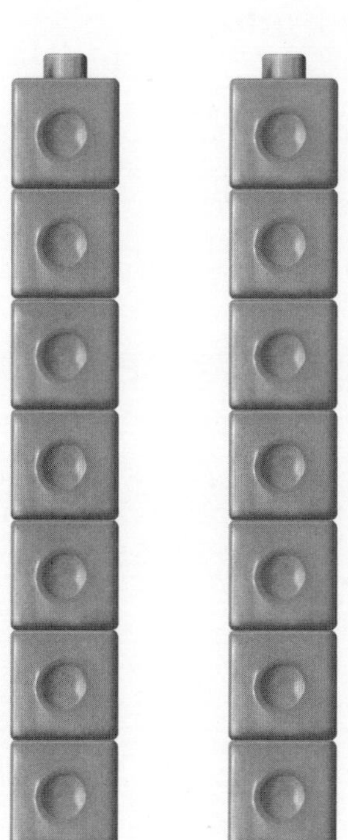

_______ + _______ = _______

4

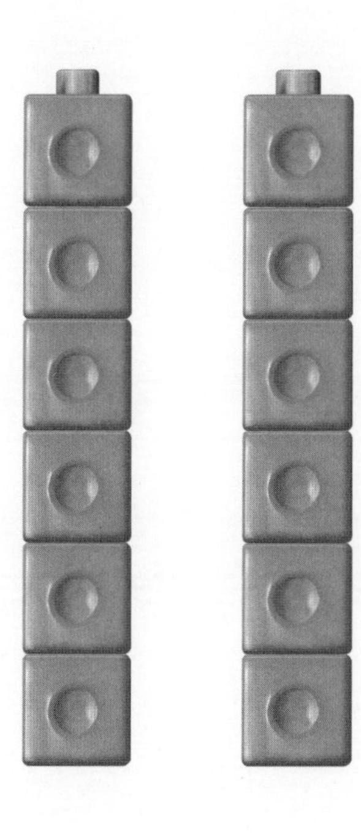

_______ + _______ = _______

☆ **Tell how all the sums are alike.**

Name ______________________________

Assessment

Circle pairs. Then circle whether the number is odd or even. Draw lines to connect pairs.

1

odd even

2

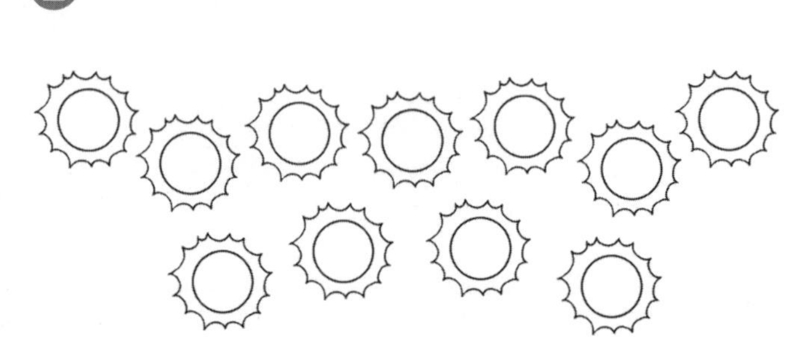

odd even

3

odd even

4 **Write the number in each tower. Then add.**

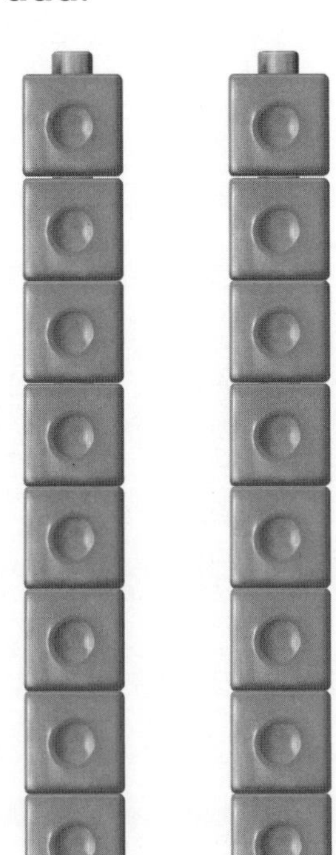

_______ + _______ = _______

☆ **Tell why an even number is the sum of two equal amounts.**

Overview Add Equal Groups

Directions and Sample Answers for Activity Pages

Day 1	See "Model the Skill" below.
Day 2	Read the directions aloud. Remind students that a column goes up and down. Explain that on this activity page, they need to count the number of footprints in each column and write the number on the line below the column. Once students have completed the page, point out that each addend is the same—3—and that they are using repeated addition to find a sum. (Answers: **1.** 3 + 3 = 6; **2.** 3 + 3 + 3 = 9; **3.** 3 + 3 + 3 + 3 = 12; **4.** 3 + 3 + 3 + 3 + 3 = 15)
Day 3	Read the directions aloud. Remind students to count only the flowers in each circled column and to write that number in the number sentence. Point out that problems 1 and 2 have operation symbols, but students will need to write the operation symbols for problems 3 and 4. (Answers: **1.** 4 + 4 = 8; **2.** 4 + 4 + 4 = 12; **3.** 4 + 4 + 4 + 4 = 16; **4.** 4 + 4 + 4 + 4 + 4 = 20)
Day 4	Read the directions aloud. Point out that there are no operation symbols shown on this page, but students should write the symbols to show how they are adding to find the sum in each problem. Remind students that the addends in the number sentences should be the number of objects in each column. Ask a volunteer to note how many moons they see in the first column of problem 1. (5) Review counting by fives from 5 through 30 with students. Tell them that they can use skip counting to help find the sum of each addition sentence. (Answers: **1.** 5 + 5 = 10; **2.** 5 + 5 + 5 = 15; **3.** 5 + 5 + 5 + 5 = 20)
Day 5	Read the directions aloud. Observe as students complete the page. Do students count the objects by columns? Do they write the plus sign to show that they are adding? Can they match an equation to an array? Use your observations to plan further instruction and review. (Answers: **1.** 3 + 3 + 3 = 9; **2.** 5 + 5 + 5 + 5 = 20; **3.** 3 + 3 + 3 = 9)

Model the Skill

- Hand out the Day 1 activity page and counters. **Say:** *A column goes up and down while a row goes across. Place a counter on each circle in the first column. How many counters did you use?* (2) Point out the number 2 at the bottom of the column. **Say:** *Now place a counter on each circle in the next column. How many counters did you use?* (2) *Look at the addition sentence below the counters. What is 2 plus 2?* (4) Observe as students write the sum of 4.
- **Say:** *The arrangement of objects in equal rows and columns is called an array. How is the array in problem 2 different than the array in problem 1?* (Possible answer: It has one more column.) Place a counter on each circle in a column and count how many counters are in each column. Check that students are placing counters in columns rather than rows. **Ask:** *What is 2 plus 2 plus 2?* (6) *Is that the total number of counters you have used?* (yes)
- Help students complete the activity page by first noting how the new array compares with that of the previous problem. Then observe students as they place counters in each column and then add to find the total number of counters. (Answers: **3.** 8; **4.** 10) **Say:** *You can use addition to find the total number of counters in an array.*

Use Manipulatives

Use counters to make arrays.

Count the number in each column.

Add the numbers to find the total number of counters.

Add Equal Groups

Place counters in each column. Add to find the total.

1.

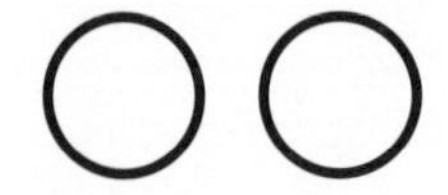

2 + 2 = ______

2.

2 + 2 + 2 = ______

3.

2 + 2 + 2 + 2 = ______

4.

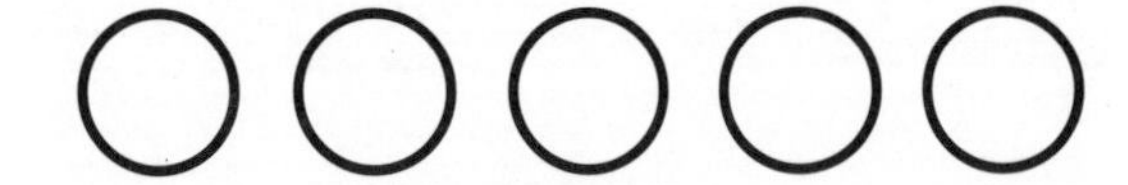

2 + 2 + 2 + 2 + 2 = ______

☆ **Tell how you can use skip counting to find the total.**

Name ____________________

Add Equal Groups

Write the number of footprints in each column. Add to find the total.

1. 

_____ + _____ = _____

2. 

_____ + _____ + _____ = _____

3. 

_____ + _____ + _____ + _____ = _____

4. 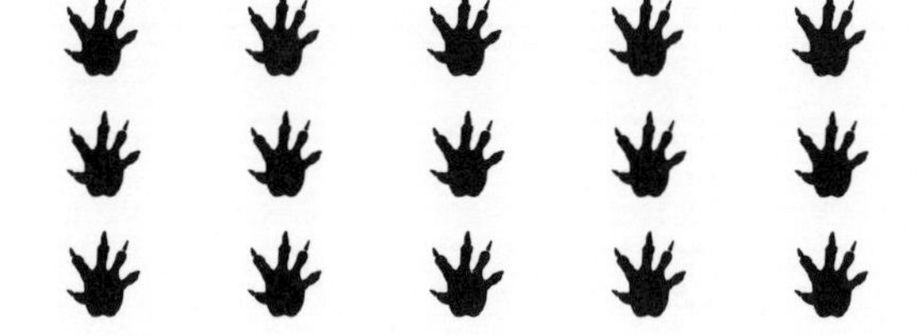

_____ + _____ + _____ + _____ + _____ = _____

☆ **Tell how you found the total in the problem.**

 Name ____________________________

Add Equal Groups

For each problem, circle the columns. Then complete the number sentence.

1

______ + ______ = ______

2

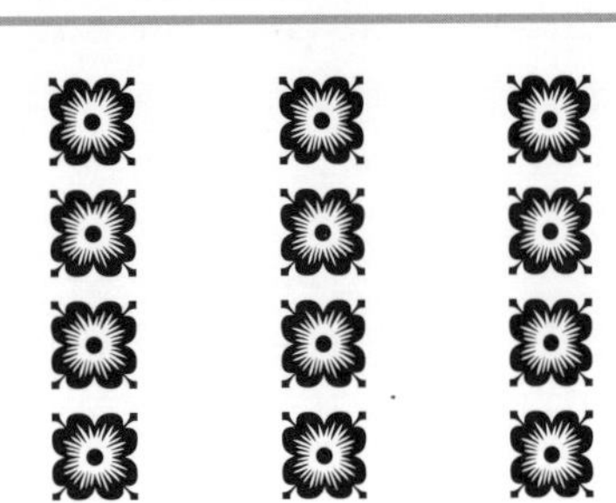

______ + ______ + ______ = ______

3

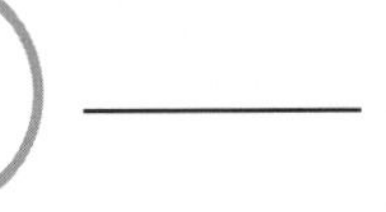

______ ◯ ______ ◯ ______ ◯ ______ = ______

4

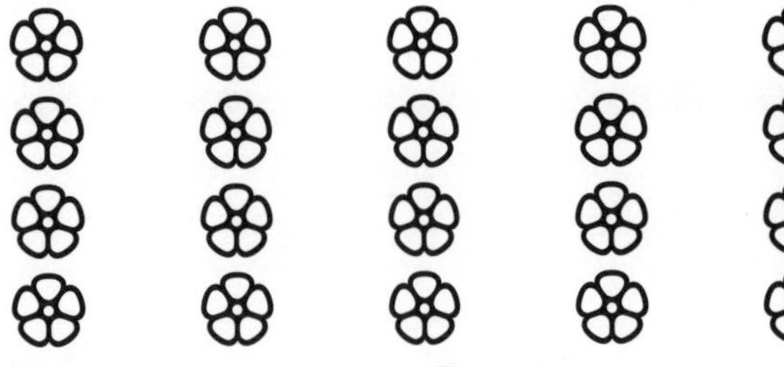

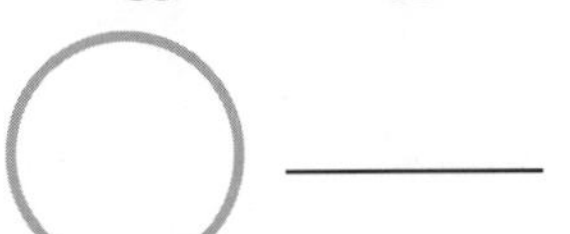

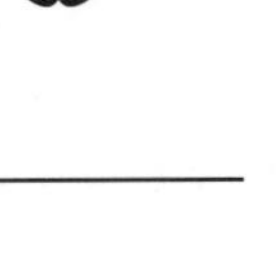

______ ◯ ______ ◯ ______ ◯ ______ ◯ = ______

☆ **Tell how you found the total number of flowers.**

 Name ______________________

Add Equal Groups

Write a number sentence for each array.

1

___ ◯ ___ ◯ ___

2

___ ◯ ___ ◯ ___ ◯ ___ ◯ ___

3

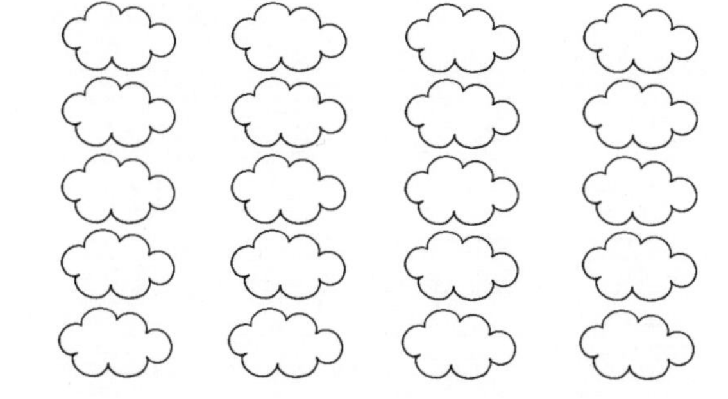

___ ◯ ___ ◯ ___ ◯ ___ ◯ ___

☆ **Tell how you can use skip counting to find the sum.**

 Name ______________________

Assessment

Write the number in each column. Then add to find the total number of objects.

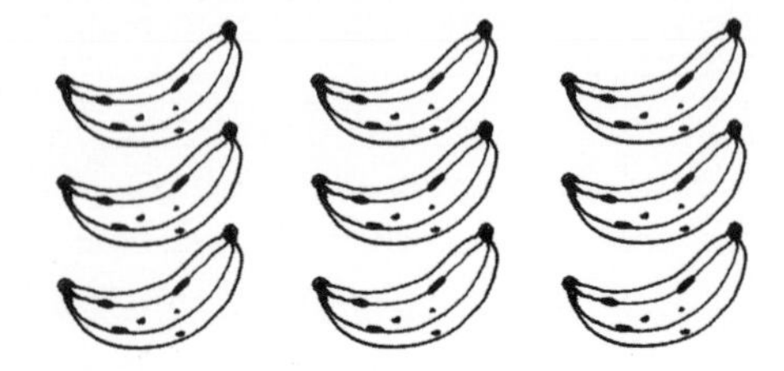

_____ + _____ + _____ = _____

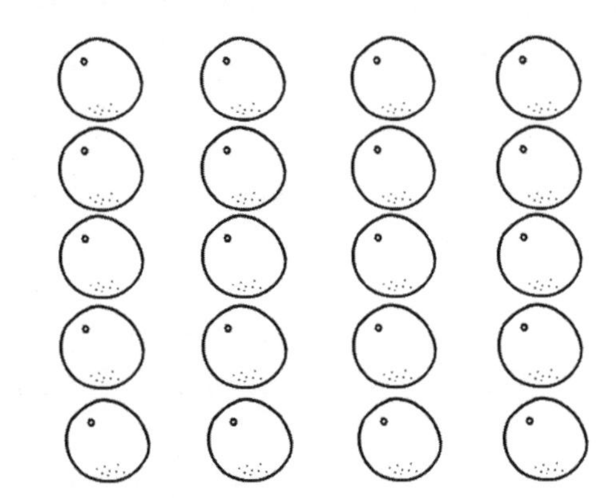

_____ ◯ _____ ◯ _____ ◯ _____ = _____

Circle the equation that shows the total number of apples.

3 + 3 + 3 = 9

4 + 4 + 4 = 12

4 + 4 + 4 + 4 = 16

☆ **Tell how you solved the problem.**

Overview Understand Place Value

Directions and Sample Answers for Activity Pages

Day 1	See "Model the Skill" below.
Day 2	Read the directions aloud. Encourage students to read the number and then to say slowly what they read as they write the numbers in the correct place. Point out that the number of hundreds they say is the number written in the hundreds place. Encourage students to check that a number is written in each place of the chart. (Answers: **1.** 5 hundreds, 6 tens, 0 ones; **2.** 4 hundreds, 0 tens, 0 ones; **3.** 7 hundreds, 3 tens, 1 one; **4.** 2 hundreds, 9 tens, 8 ones)
Day 3	Read the directions aloud. Have students look at each place in the chart and transfer the amounts to the expanded notation sentence format. Remind students that there should be an amount indicated for each place. Have students transfer the amounts from the expanded form to the standard form. (Answers: **1.** 3 hundreds, 0 tens, 8 ones = 308; **2.** 8 hundreds, 2 tens, 6 ones = 826; **3.** 1 hundred, 5 tens, 7 ones = 157; **4.** 6 hundreds, 0 tens, 0 ones = 600)
Day 4	Read the directions aloud. Point out that there are no place-value charts on this form. Remind students that a three-digit number tells the number of hundreds, the number of tens, and the number of ones. (Answers: **1.** 4 hundreds, 2 tens, 0 ones; **2.** 2 hundreds, 9 tens, 3 ones; **3.** 7 hundreds, 0 tens, 0 ones; **4.** 5 hundreds, 8 tens, 6 ones)
Day 5	Read the directions aloud. Observe as students complete the page. Do students write a number in each place? Do they remember to use zero when there are no tens or ones? (Answers: **1.** 1 hundred, 7 tens, 5 ones; **2.** 9 hundreds, 0 tens, 0 ones; **3.** 3 hundreds, 2 tens, 8 ones; **4.** 6 hundreds, 0 tens, 3 ones)

Model the Skill

- Hand out the Day 1 activity page and base-ten blocks. **Say:** *A number can be shown with base-ten blocks.* Review with students that 10 ones equal 1 ten. Then show 10 tens. **Ask:** *What number is shown with 10 tens?* (100) Have students count the rods by tens to 100. **Ask:** *What number is written in problem 1?* (234) *Use blocks to show that number. How many hundreds do you use?* (2) *How many tens?* (3) *How many ones?* (4) Observe as students write the number of hundreds, tens, and ones below the pictured blocks. (2 hundreds, 3 tens, 4 ones)

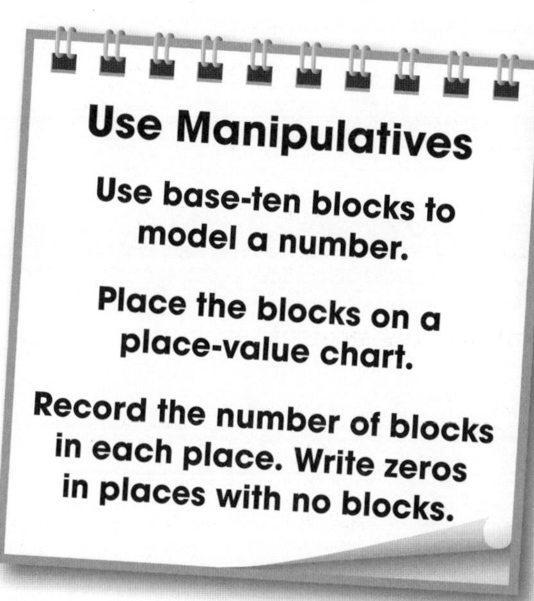

- **Ask:** *How do you show the number in problem 2?* (Possible answer: with 3 hundreds) *Why are there no tens or ones shown in the chart?* (Possible answer: There are no tens or ones in the number 300.) Write the numbers **3** and **300** on the board. **Say:** *Even though only 3 hundreds are needed to show the number, you need to write the zeros in the tens and ones places to indicate that no tens and ones are needed.* Have students write the numbers in the chart. (3 hundreds, 0 tens, 0 ones)

- Help students complete the activity page by showing each number with base-ten blocks and recording the number of blocks in each place. Remind students to use zeros in places where there are no blocks. Have students compare the written three-digit numbers with the numbers written in the charts. (Answers: **3.** 2 hundreds, 0 tens, 9 ones; **4.** 1 hundred, 5 tens, 7 ones)

Understand Place Value

Show each number with base-ten blocks. Write the number of hundreds, tens, and ones.

1. **234**

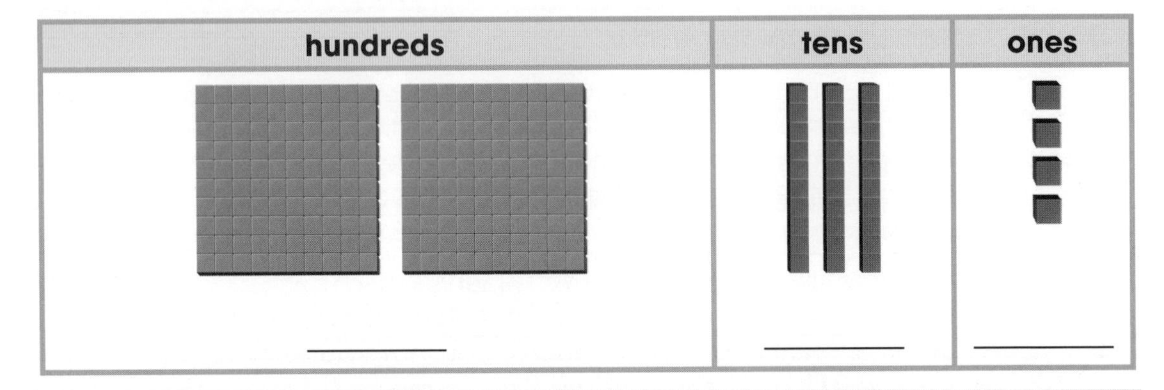

2. **300**

hundreds	tens	ones
______	______	______

3. **209**

hundreds	tens	ones
______	______	______

4. **157**

hundreds	tens	ones
______	______	______

☆ **Tell how you know what to write in each column.**

Understand Place Value

Write the number in the correct places of the chart.

1. **560**

hundreds	tens	ones

2. **400**

hundreds	tens	ones

3. **731**

hundreds	tens	ones

4. **298**

hundreds	tens	ones

☆ **Tell how you know what number to write in each place.**

 Name ____________________

Understand Place Value

Write the number of hundreds, tens, and ones. Then write the number.

hundreds	tens	ones
3	0	8

____ hundreds ____ tens ____ ones = ____

hundreds	tens	ones
8	2	6

____ hundreds ____ tens ____ ones = ____

hundreds	tens	ones
1	5	7

____ hundreds ____ tens ____ ones = ____

hundreds	tens	ones
6	0	0

____ hundreds ____ tens ____ ones = ____

☆ **Tell how you know the number of tens.**

 Name ______________________________

Understand Place Value

Write the number of hundreds, tens, and ones.

420 = _______ hundreds _______ tens _______ ones

293 = _______ hundreds _______ tens _______ ones

700 = _______ hundreds _______ tens _______ ones

586 = _______ hundreds _______ tens _______ ones

☆ **Tell how you know the number of hundreds.**

 Name ____________________

Assessment

Write the number of hundreds, tens, and ones.

175

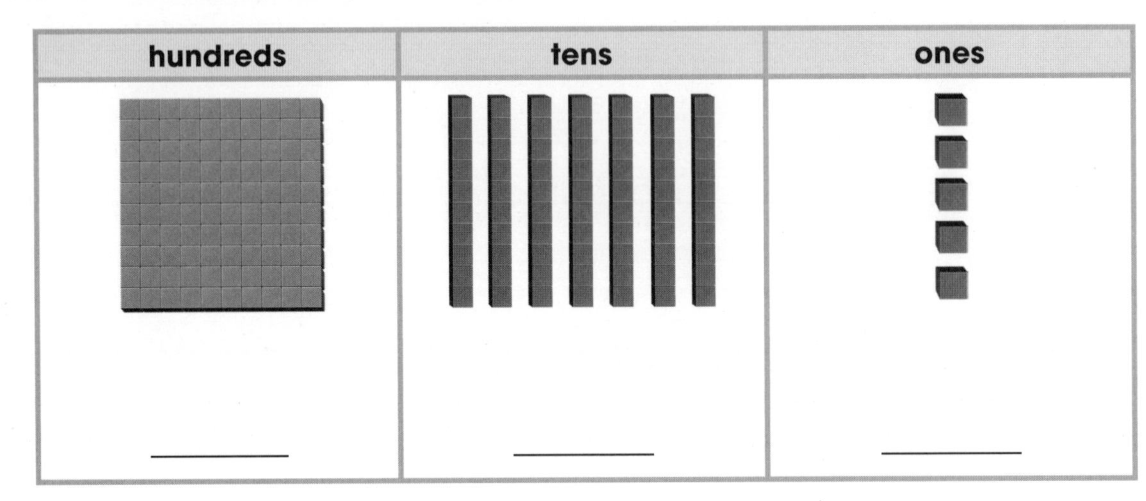

hundreds	tens	ones
____	____	____

hundreds	tens	ones
9	0	0

____ hundreds ____ tens ____ ones = ____

328 = ______ hundreds ______ tens ______ ones

4

603 = ______ hundreds ______ tens ______ ones

☆ **Tell why you need to write a zero in the number.**

Overview Count, Read, and Write Numbers to 1,000

Directions and Sample Answers for Activity Pages

Day 1	See "Model the Skill" below.
Day 2	Read the directions aloud. Remind students that the expanded form of a number shows the number as a sum of the hundreds, tens, and ones. Explain that the standard form uses just the numerals. Encourage students to say the number as they write the number words. (Answers: **1.** 170; 100 + 70; one hundred seventy; **2.** 202; 200 + 2; two hundred two; **3.** 364; 300 + 60 + 4; three hundred sixty-four)
Day 3	Read the directions aloud. Tell students to skip count by 100s for problems 1 and 2, and skip count by 10s for problems 3 and 4. Review counting by 10s to 100. (Answers: **1.** 400, 500, 600, 700; **2.** 600, 700, 800, 900; **3.** 440, 450, 460, 470; **4.** 880, 890, 900, 910)
Day 4	Read the directions aloud. Explain that students should skip count by 5s for problems 1 and 2, and then use skip counting to find the missing numbers in problems 3 and 4. Review counting by 5s to 100. (Answers: **1.** 525, 530, 535, 540; **2.** 790, 795, 800, 805; **3.** 660, 670, 680; **4.** 960, 965, 970)
Day 5	Read the directions aloud. Observe as students complete the page. Do students understand each representation? Do they identify the correct skip-counting pattern? Use your observations to plan further instruction and review. (Answers: **1.** 700 + 500 + 9; **2.** 8 hundreds 4 tens; **3.** 130, 135, 140; **4.** 800, 900, 1,000)

Model the Skill

- Hand out the Day 1 activity page and base-ten blocks. **Say:** *Today we will show the same number in different ways.* Read aloud the number in problem 1. (three hundred twenty-four) Show that number with base-ten blocks. Have students build the number with blocks. Guide them to write the number of hundreds, tens, and ones.

- **Say:** *You can show a number as a sum of the hundreds, tens, and ones. Look at the third column. What would you write to show the numeral for 3 hundreds?* (a 3 in front of the zeros) *What do you need to write to show the numeral for 2 tens?* (a 2 in front of the zero) *What do you need to write to show the numeral for 4 ones?* (a 4) Refer to all three columns to show the same number in different ways.

- Have students read aloud the number in problem 2, show it with base-ten blocks, and complete the second column of the chart. **Say:** *This number does not have any tens. We don't write zeros by themselves when writing a number in expanded form. We just write the places that have numbers. How can you show the number of hundreds?* (with a 6 in front of the zeros to show 600) *How can you show the number of ones?* (with a 9) *So the expanded form of 609 is 600 + 9.*

- Help students complete the activity page, noting that they need to write the zeros for the hundreds as well as the tens in the expanded form. (Answers: **3.** 4 hundreds, 8 tens, 0 ones; 400 + 80; **4.** 2 hundreds, 5 tens, 1 one; 200 + 50 + 1)

Use Counting

Practice counting to 1,000 throughout the day with or without objects.

Count by 1s starting at various numbers.

Then count by 5s, 10s, or 100s starting at various numbers.

 Name ____________________

Count, Read, and Write Numbers to 1,000

Read the number. Show it with base-ten blocks. Complete the chart.

Number	Place Value
324	_____ hundreds _____ tens _____ ones **Expanded Form:** _____**00** + _____**0** + _____
609	_____ hundreds _____ tens _____ ones **Expanded Form:** _____**00** + _____
480	_____ hundreds _____ tens _____ ones **Expanded Form:** _____ + _____ + _____
251	_____ hundreds _____ tens _____ ones **Expanded Form:** _____ + _____ + _____

☆ **Tell how you know how to read the number.**

Count, Read, and Write Numbers to 1,000

Write the number shown in three different ways. Use base-ten blocks if you wish.

1

hundreds	tens	ones
1	7	0

standard form 170

expanded form 100 + ____________

word form ____________

hundreds	tens	ones
2	0	2

standard form ____________

expanded form ____________

word form ____________

hundreds	tens	ones
3	6	4

standard form ____________

expanded form ____________

word form ____________

☆ **Tell how you know the different ways to write a number.**

Count, Read, and Write Numbers to 1,000

Skip count by 100s. Continue counting from the numbers given.

1

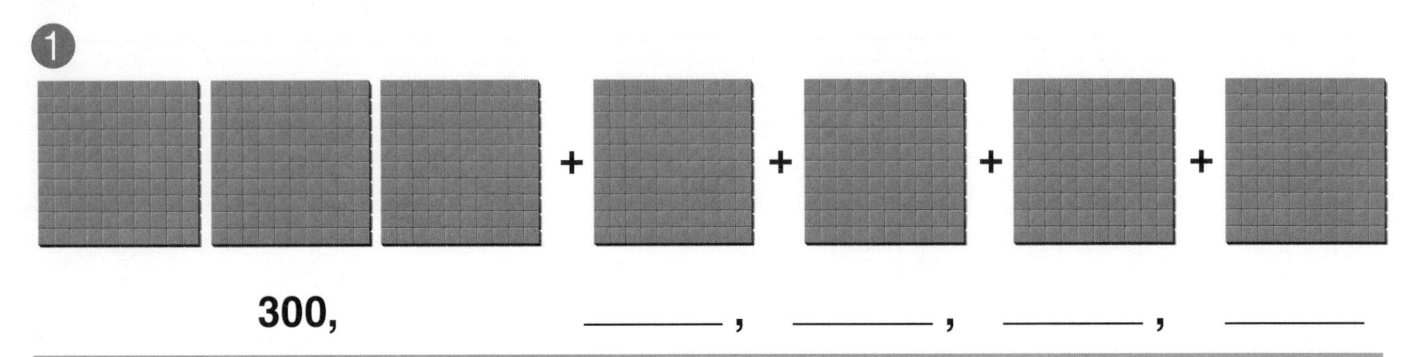

300, ______, ______, ______, ______

2

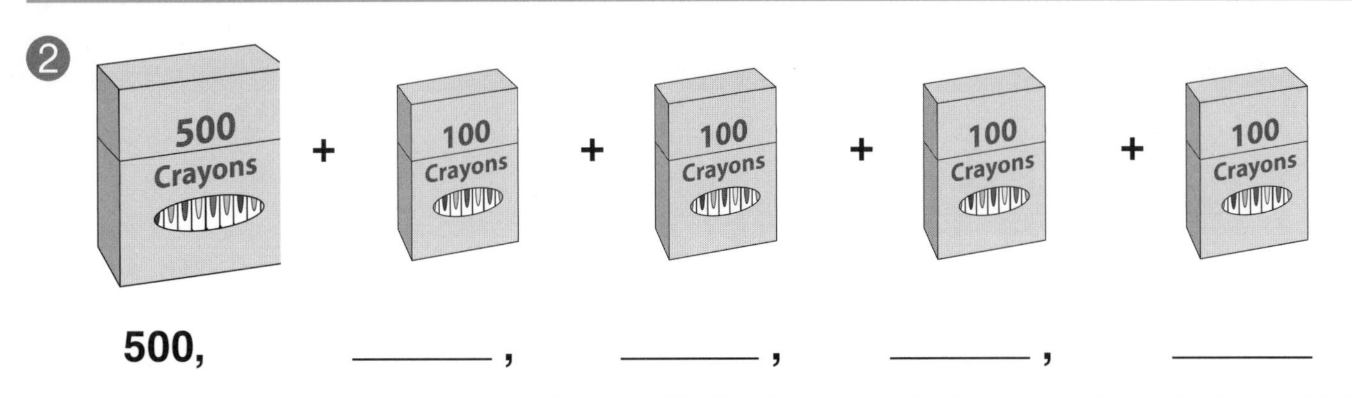

500, ______, ______, ______, ______

Skip count by 10s. Continue counting from the numbers given.

3

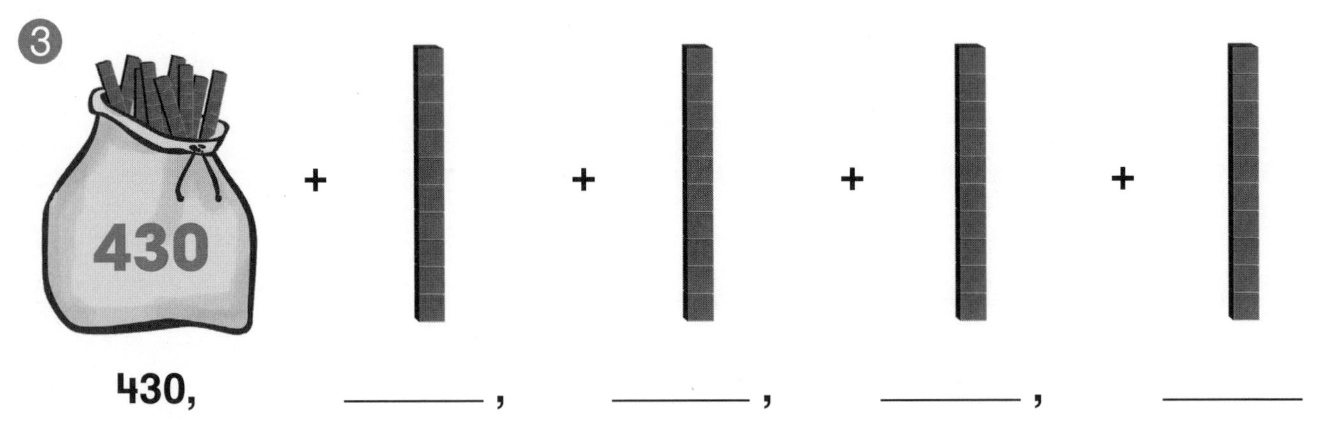

430, ______, ______, ______, ______

4

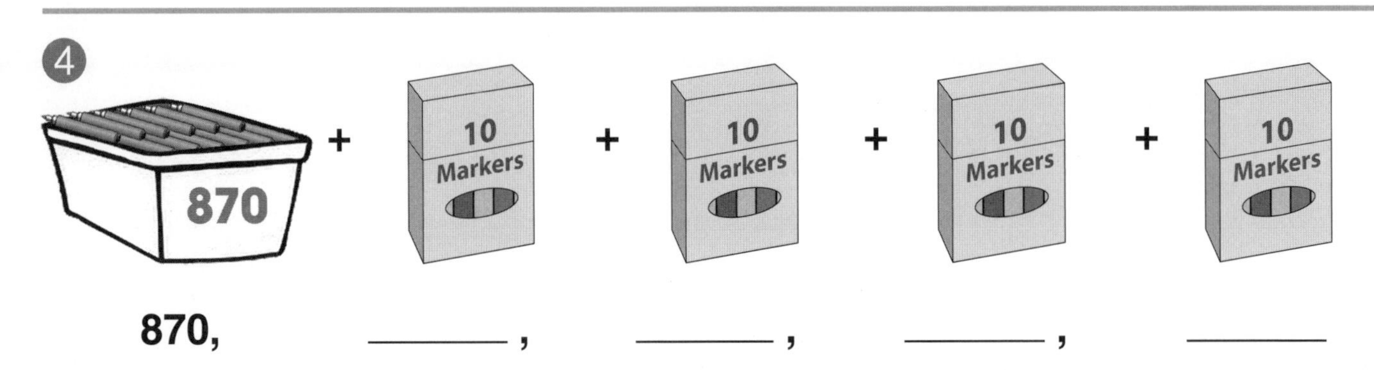

870, ______, ______, ______, ______

☆ **Tell how you know which number comes next.**

Count, Read, and Write Numbers to 1,000

Skip count by 5s. Continue counting from the numbers given.

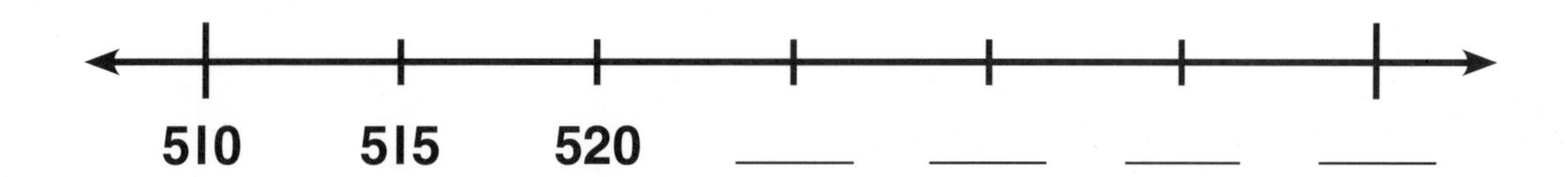

2

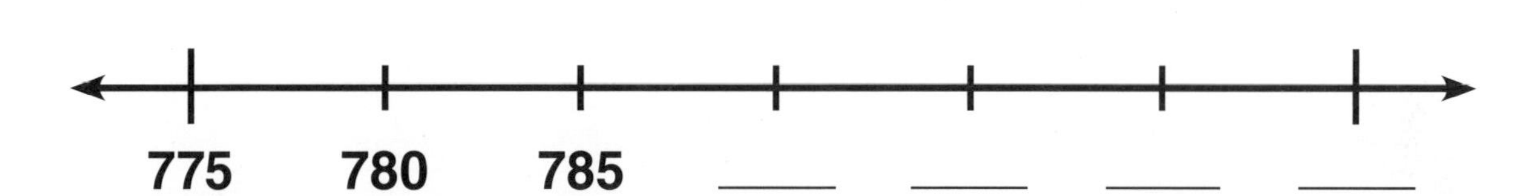

Look for a skip-counting pattern. Write the missing numbers.

630, 640, 650, ______, ______, ______

4

945, 950, 955, ______, ______, ______

☆ **Tell how you know which number comes next when skip counting by 5s.**

Assessment

Cross out the number that does not match.

1

759

seven hundred fifty-nine

7 hundreds **5** tens **9** ones

700 + 500 + 9

2

eight hundred four

800 + 4

804

8 hundreds 4 tens

Look for a skip-counting pattern. Write the missing numbers.

3

115, 120, 125, _______, _______, _______

4

500, 600, 700, _______, _______, _______

☆ **Tell how you know what comes next.**

Overview Compare Numbers

Directions and Sample Answers for Activity Pages

Day 1	See “Model the Skill” below.
Day 2	Read the directions aloud. Tell students that they will compare the numbers of hundreds, tens, and ones in the two numbers. Explain that when they find an unmatched block, it indicates that one number is greater than the other. Discuss the different meanings of the comparison statements that students will circle. (Answers: **1.** 109 is greater than 107. **2.** 215 is greater than 213. **3.** 115 is less than 125. **4.** 224 is equal to 224.)
Day 3	Read the directions aloud. Refer students to the box that shows the meaning of each comparison symbol. Point out that the wider side of the symbol should be on the side with the greater number, and the smaller side of the symbol should be on the side with the lesser number. You may also tell students that they can think of the “greater than” and “less than” symbols as a mouth that eats the greater number. Remind students to compare the hundreds before comparing the tens or the ones. (Answers: **1.** 845 > 841; **2.** 739 < 779; **3.** 654 < 954; **4.** 502 = 502)
Day 4	Read the directions aloud. Review the meanings of the comparison symbols with students. Tell students that they should place their pencil closer to the greater number and move it toward the lesser number as they write the symbol inside the circle. (Answers: **1.** >; **2.** =; **3.** <; **4.** >)
Day 5	Read the directions aloud. Observe as students complete the page. Do students compare the hundreds before they compare the tens or ones? Do they write the correct comparison symbol? Use your observations to plan further instruction and review. (Answers: **1.** 135 is less than 153. **2.** >; **3.** =; **4.** <)

Model the Skill

- Hand out the Day 1 activity page and base-ten blocks. **Say:** *Let's compare two numbers. Show the two numbers in problem 1 with the base-ten blocks.* Show the numbers in separate locations. **Say:** *First, you need to compare the greatest place—the hundreds. Does one number have more hundreds than the other?* (no) *Compare the next place—the tens. Does one number have more tens than the other?* (Yes, 132 has more tens.) Have students circle 132 on the activity page. Point out that there is no need to compare the ones because they have already determined which number is greater. **Say:** *132 is greater than 123 and 123 is less than 132.*
- **Say:** *Show the two numbers in problem 2 with blocks.* Point to a group that shows 205. **Say:** *This group must be greater because it has more blocks. Am I right?* (no) *Why not?* (Possible answer: Having more blocks doesn't mean the number is greater; you have to look at the values of the blocks.) Have students compare the hundreds and then the tens. Point out that 205 does not have any tens while 210 has 1 ten. **Say:** *1 ten is equal to 10 ones. If you showed 210 with 2 hundreds and 10 ones, there would be more blocks.* Have students circle 210 on the activity page. **Say:** *210 is greater than 205 and 205 is less than 210.*
- Help students complete the activity page by showing the numbers with base-ten blocks, comparing the places, and circling the greater number. (Answers: **3.** 237; **4.** 228)

Use Manipulatives

Use base-ten blocks to model the two numbers.

Compare the hundreds. If one number has more, it is greater.

If the hundreds are the same, compare the tens. Then compare the ones.

Compare Numbers

Show each number with blocks. Compare. Circle the greater number.

1

123 132

2

210 205

3

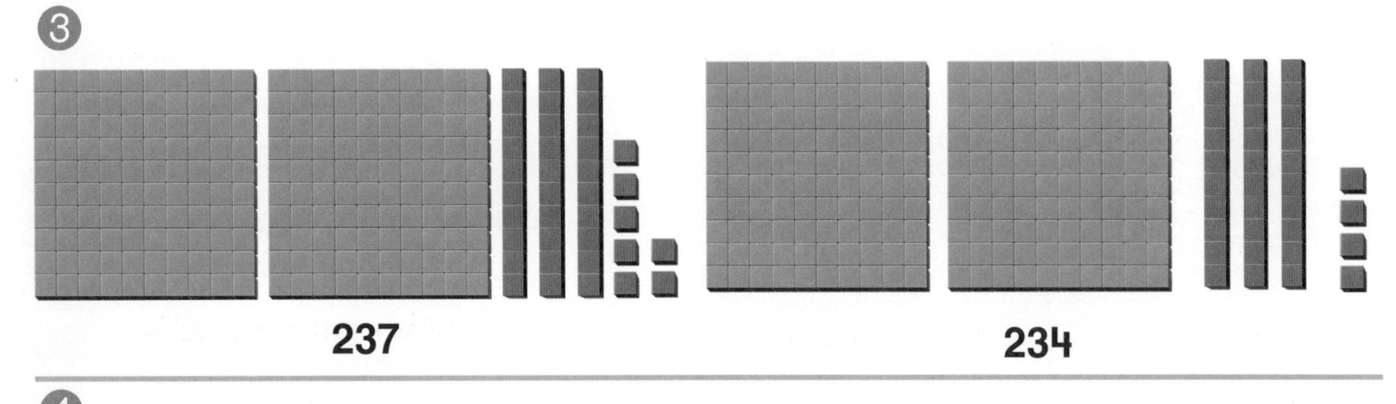

237 234

4

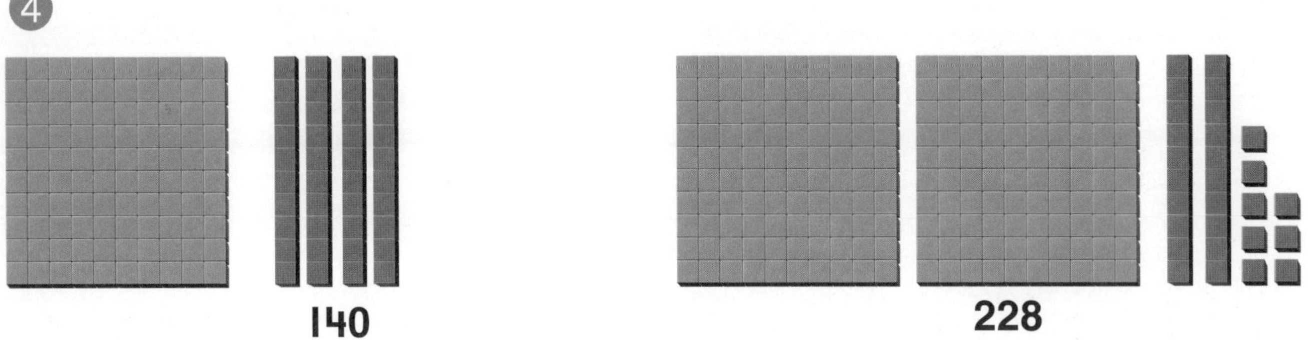

140 228

☆ **Tell how you know which number is greater.**

Compare Numbers

Compare each number block. Circle the true statement.

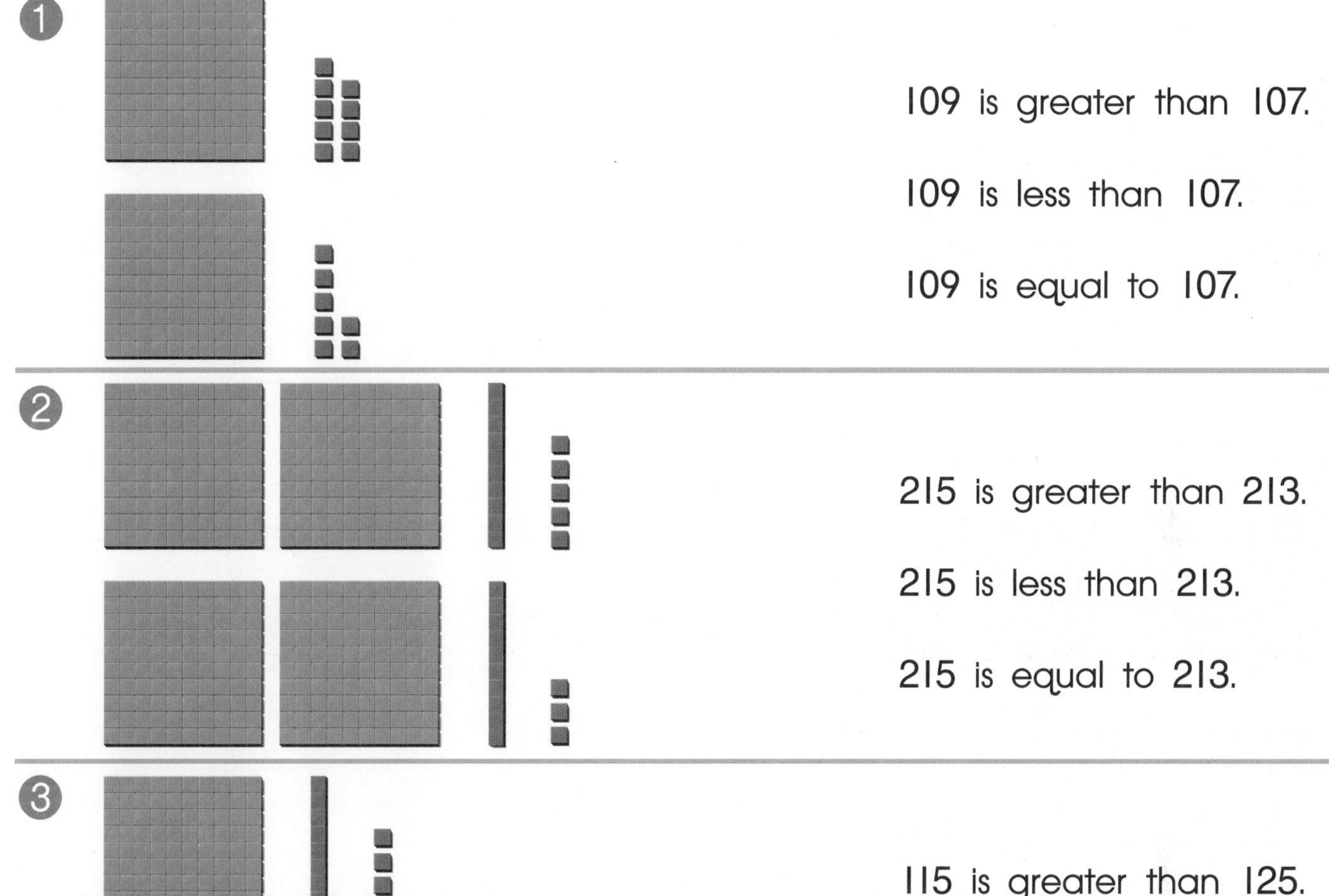

1.

109 is greater than 107.

109 is less than 107.

109 is equal to 107.

2.

215 is greater than 213.

215 is less than 213.

215 is equal to 213.

3.

115 is greater than 125.

115 is less than 125.

115 is equal to 125.

4.

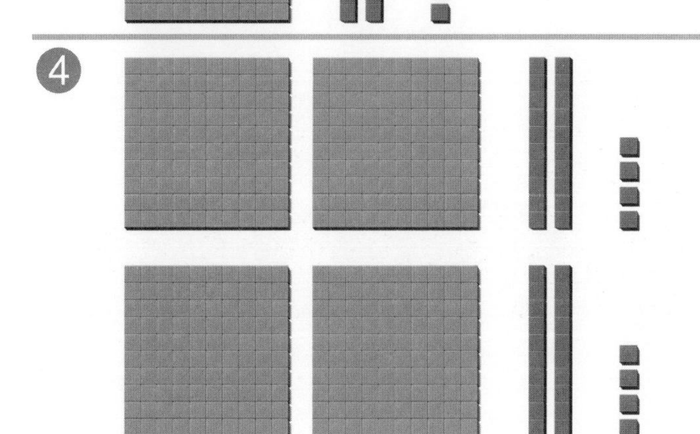

224 is greater than 224

224 is less than 224.

224 is equal to 224.

☆ **Tell how you know when a number is less than the other number.**

Compare Numbers

Compare. Circle >, <, or = for each problem.

> is greater than; < is less than; = is equal to

1

hundreds	tens	ones
8	4	5
8	4	1

>
845 < 841
=

2

hundreds	tens	ones
7	3	9
7	7	9

>
739 < 779
=

3

hundreds	tens	ones
6	5	4
9	5	4

>
654 < 954
=

4

hundreds	tens	ones
5	0	2
5	0	2

>
502 < 502
=

☆ **Tell how you know which symbol to circle.**

Compare Numbers

Write >, <, or = to compare.

> is greater than; < is less than; = is equal to

1

hundreds	tens	ones
4	2	6
4	0	9

426 ◯ 409

2

hundreds	tens	ones
6	5	8
6	5	8

658 ◯ 658

2

934 ◯ 943

4

720 ◯ 270

☆ **Tell how you know which symbol to write.**

Assessment

Match to compare. Circle the true comparison statement.

1.

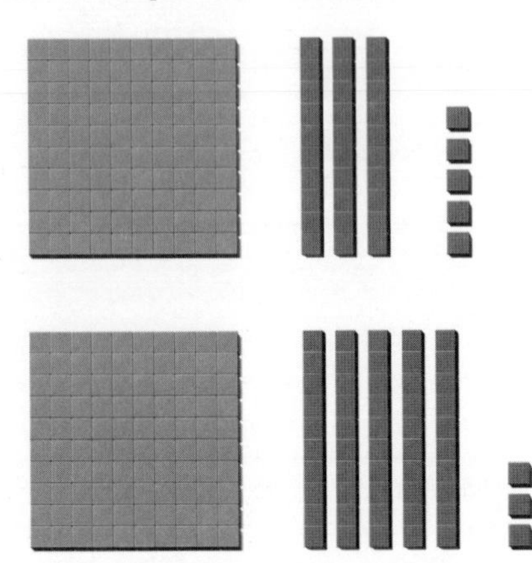

135 is greater than 153.

135 is less than 153.

135 is equal to 153.

> is greater than; < is less than; = is equal to

2.

hundreds	tens	ones
6	0	8
6	0	0

608 ◯ 600

3.

345 ◯ 345

4.

709 ◯ 790

☆ **Tell how you solved the problem.**

Overview Use Strategies to Add

Directions and Sample Answers for Activity Pages

Day 1	See "Model the Skill" below.
Day 2	Read the directions aloud. Explain to students that they can add numbers in any order and the sum will not change. Point out that for problems 1 and 2, students can begin with the second addend and count on to find the sum. For problems 3 and 4, students should look for two addends with ones that equal 10. Then they should add the sum of those addends to the third addend. Direct students to show their work in the boxes on the page. (Answers: **1.** 31; **2.** 38; **3.** 52; **4.** 61)
Day 3	Read the directions aloud. Tell students that another addition strategy is to use models, such as base-ten blocks. Tell students that they should add the ones blocks and then add the tens blocks. Remind students that when there are more than 10 ones blocks, they will need to regroup 10 ones as 1 ten. Point out that regrouping is needed in problems 3 and 4. (Answers: **1.** 50; **2.** 39; **3.** 31; **4.** 21)
Day 4	Read the directions aloud. Tell students that a place-value chart can be used to help them focus on adding the numerals in each place. Point out that the equation has been written in the place-value chart. Remind students to add the ones before adding the tens and to regroup when there are more than 10 ones. Regrouping is needed in problems 3 and 4. (Answers: **1.** 54; **2.** 39; **3.** 53; **4.** 71)
Day 5	Read the directions aloud. Observe as students complete the page. Do students look for addends that equal a decade number when adding three numbers? Do they remember to regroup when needed? Do they add the ones before the tens? Use your observations to plan further instruction and review. (Answers: **1.** 41; **2.** 47; **3.** 34; **4.** 48)

Model the Skill

- Hand out the Day 1 activity page.
- **Say:** *You can use different strategies to add. You can count on by 1, 2, or 3. A number line can help you.* Have students look at problem 1 and circle the first addend on the number line. **Say:** *You can draw two jumps to count on 2 from 25.* Show students how to draw a curved line from 25 to 26 and then from 26 to 27 to show two jumps. **Ask:** *On what number did you land?* (27)
- Have students look at problem 2. **Ask:** *What number will you circle on the number line?* (31) *How many jumps will you make?* (3) *What is 31 plus 3?* (34)
- Help students complete the activity page by counting on to add on the number line. Remind students to circle the first addend and draw jumps to the right on the number line equal to the second addend. (Answers: **3.** 32; **4.** 50)

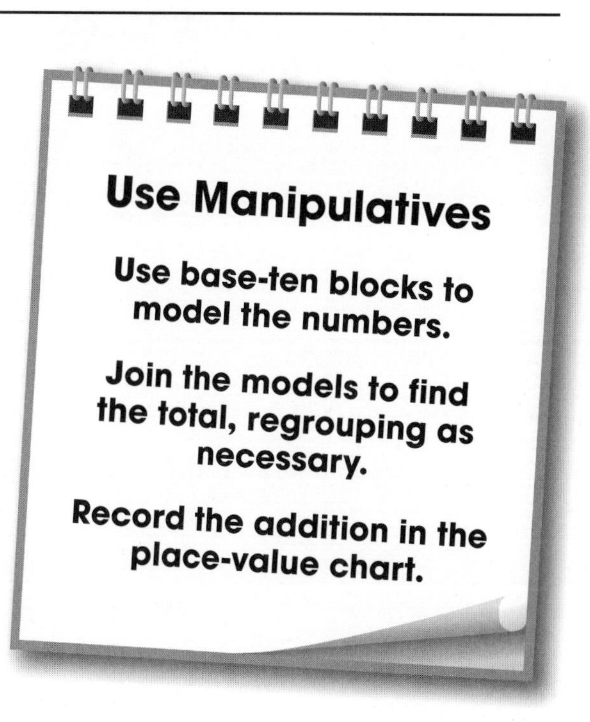

Use Strategies to Add

Count on to add. Write each sum.

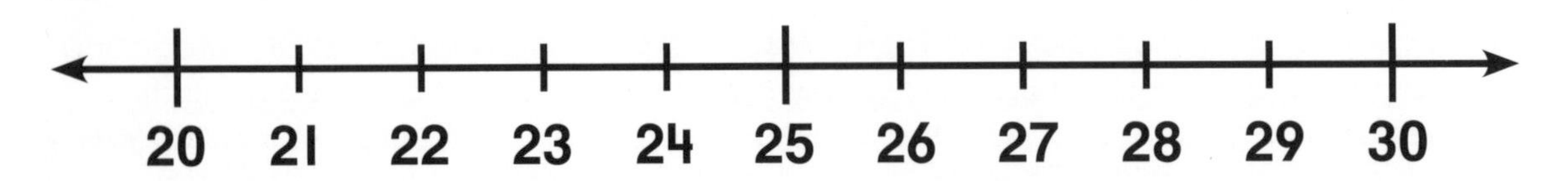

25 + 2 = ______________

2

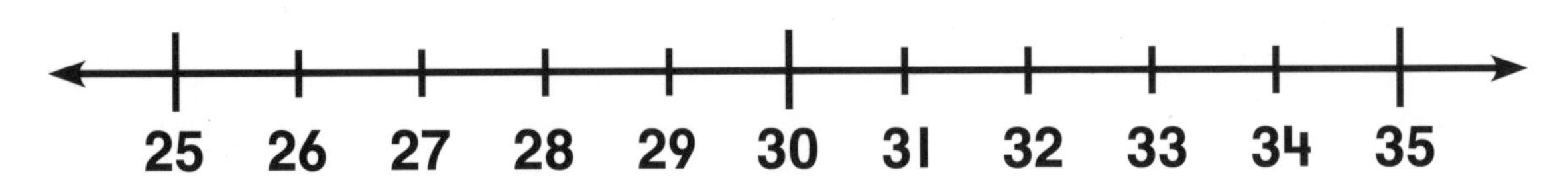

31 + 3 = ______________

3

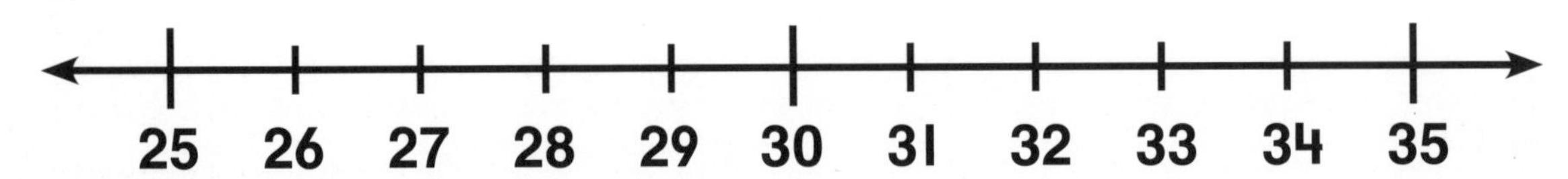

29 + 3 = ______________

4

45 46 47 48 49 50 51 52 53 54 55

49 + 1 = ______________

☆ **Tell how you can use a number line to add.**

Name ____________________

Use Strategies to Add

Add in any order. Show your thinking. Write each sum.

3 + 28 = ________

2 + 36 = ________

Circle the numbers you will add first. Show your work. Write the sum.

32 + 15 + 5 = ________

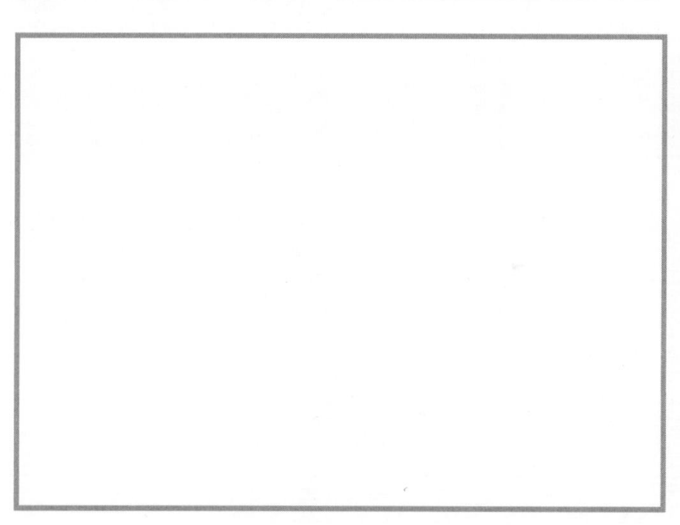

7 + 41 + 13 = ________

☆ **Tell how you solved the problem.**

Name ______________________

Use Strategies to Add

Find each sum.

1

41 + 9 = ______________

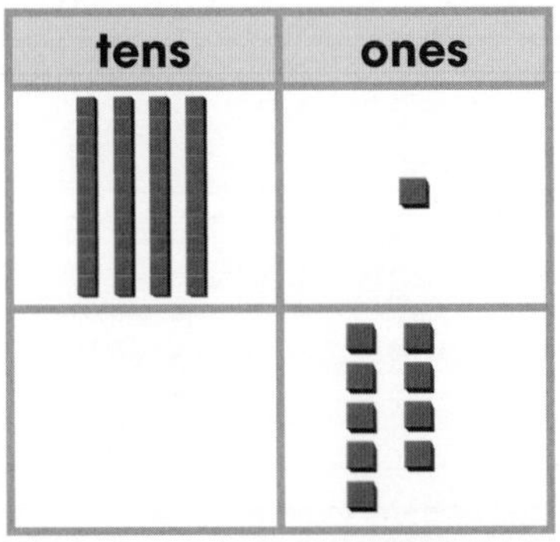

	tens	ones
	4 tens	1 one
+		9 ones

2

9 + 30 = ______________

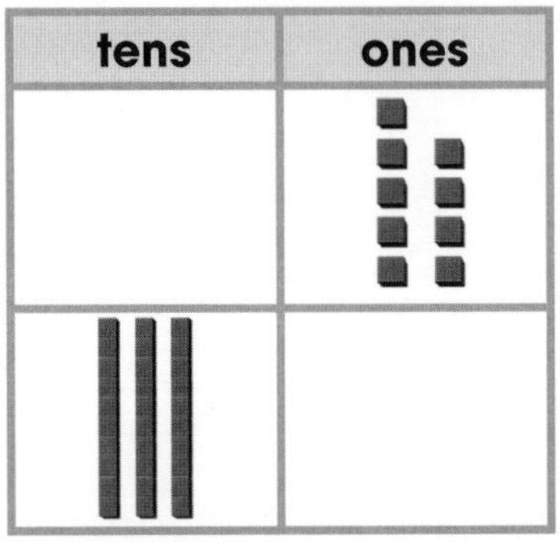

	tens	ones
		9 ones
+	3 tens	

3

8 + 23 = ______________

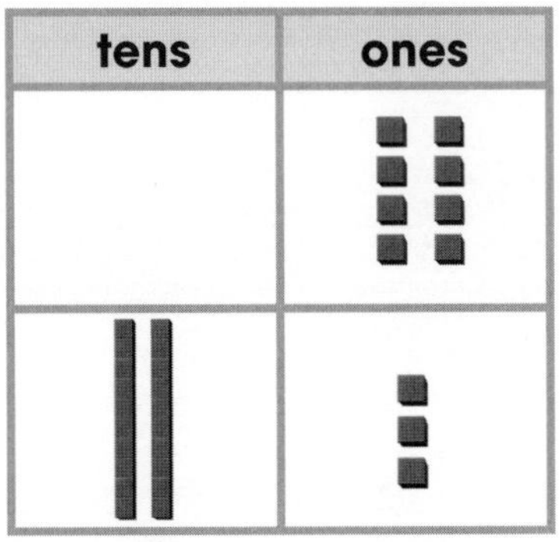

	tens	ones
		8 ones
+	2 tens	3 ones

4

16 + 5 = ______________

	tens	ones
	1 ten	6 ones
+		5 ones

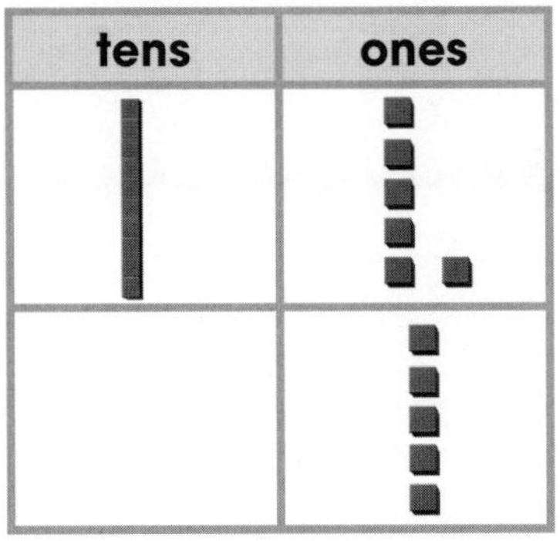

☆ **Tell how you used blocks to add.**

Use Strategies to Add

Find each sum.

1

50 + 4

	tens	ones
	5	0
+		4

2

6 + 33

	tens	ones
		6
+	3	3

3

45 + 8

	tens	ones
	4	5
+		8

4

7 + 64

	tens	ones
		7
+	6	4

☆ **Tell how you used place value to add.**

 Name ______________________________

Assessment

Find each sum. Show your thinking.

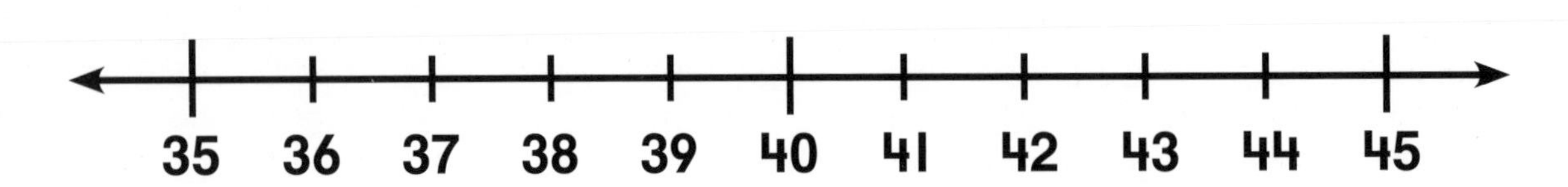

38 + 3 = ______________________

2

16 + 27 + 4 = ______________________

3

25 + 9 = ______________________

	tens	ones
+		

8 + 40

	tens	ones
		8
+	4	0

☆ **Tell how you solved the problem.**

Overview Add Two-Digit Numbers

Directions and Sample Answers for Activity Pages

Day 1	See "Model the Skill" below.
Day 2	Read the directions aloud. Explain to students that base-ten blocks can help them add two numbers. Point out how the blocks match the addends in the horizontal equation. Tell students that writing the addition in vertical form in a place-value chart can help them add the ones and then the tens. Point out that students are to write the addends in the place-value chart in problems 3 and 4. Observe that students add the ones before they add the tens. (Answers: **1.** 65; **2.** 73; **3.** 59; **4.** 49)
Day 3	Read the directions aloud. Tell students that when they add the ones, they will have more than 10; so, they will need to regroup 10 ones as 1 ten. Guide students through problem 1 by adding the ones, writing 2 ones at the bottom of the ones column and recording 1 ten at the top of the tens column. Point out that students are to write the addends in the place-value chart in problems 3 and 4. Check that students regroup and include the regrouped ten in each sum. (Answers: **1.** 72; **2.** 66; **3.** 62; **4.** 73)
Day 4	Read the directions aloud. Remind students that they can add numbers in any order and the sum will not change. Guide students to look for two addends with ones that equal 10 and to write those addends in the first vertical frame. Point out that an arrow guides them to write the sum of the two addends as the first addend in the second frame and then to add the third addend. Students should continue to add the ones before adding the tens. (Answers: **1.** 86; **2.** 93; **3.** 94; **4.** 81)
Day 5	Read the directions aloud. Have students complete the page. Do students add the ones before the tens? Do they look for a sum of ten ones when choosing which two of three addends to add? Use your observations to plan further instruction. (Answers: **1.** 50 + 6 = 56; **2.** 74; **3.** 79; **4.** 95)

Model the Skill

- Hand out the Day 1 activity page and base-ten blocks. Have students show the two numbers in problem 1 with base-ten blocks. **Say:** *Let's add these amounts by first joining the tens. How many tens are there in all?* (5) Point out the equation 20 + 30. Have students write the sum. (50) **Say:** *Let's add the ones. How many ones are there in all?* (5) Point out the equation 1 + 4. Have students write the sum. (5) Have them write the two partial sums in the vertical frame—50 + 5. **Say:** *Now add the sum of the tens and the sum of the ones. What is the total sum?* (55)
- Have students look at problem 2. **Ask:** *Where did the equation 30 + 10 come from?* (Possible answer: It is the addition of the tens in 35 + 12.) *What shows the addition of the ones?* (5 + 2) Have students show the numbers with blocks, add the tens (40), and add the ones. (7) **Say:** *Once you have found the sum of the tens and ones, what do you do?* (Add the tens and ones together.) Write the partial sums and calculate the total sum. (47)
- Help students complete the activity page by modeling the addends, adding the tens, adding the ones, and writing the partial sums in the vertical frame. Point out that the sum of the ones is a two-digit number. Have them add the ones and the tens when finding the final sum. (Answers: **3.** 50 + 11 = 61; **4.** 40 + 15 = 55)

Use Manipulatives

Use base-ten blocks to model two numbers.

Join the models to find the total, regrouping as necessary.

Record the addition in both horizontal (equation) and vertical formats.

 Name ____________________

Add Two-Digit Numbers

Show each addend with blocks. Add the tens. Add the ones. Add the sums.

1

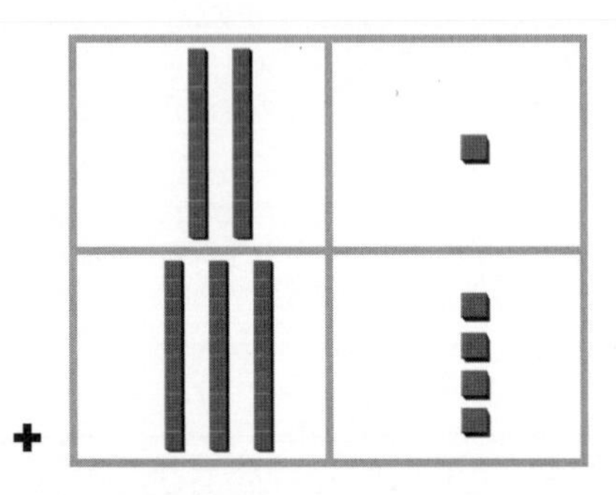

21 + 34

20 + 30 = ______ ⇨ ______

1 + 4 = ______ ⇨ ______

2

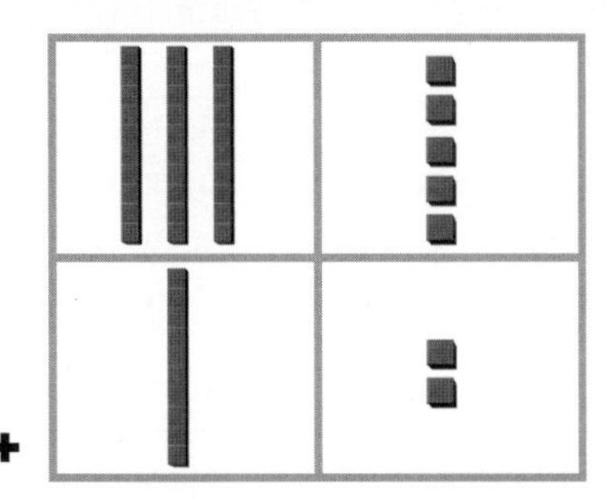

35 + 12

30 + 10 = ______ ⇨ ______

5 + 2 = ______ ⇨ ______

3

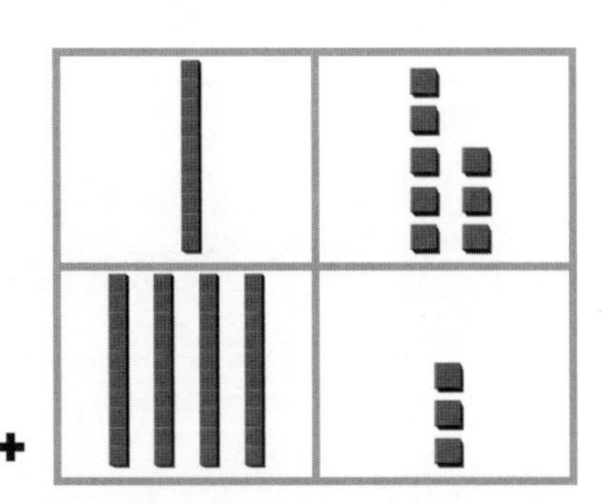

18 + 43

10 + 40 = ______ ⇨ ______

8 + 3 = ______ ⇨ ______

4

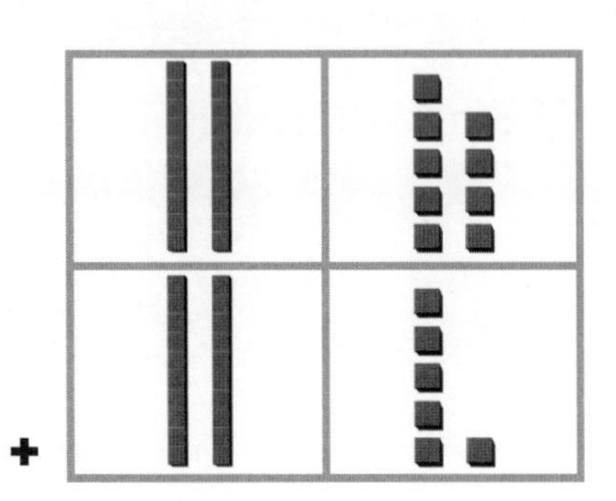

29 + 26

20 + 20 = ______ ⇨ ______

9 + 6 = ______ ⇨ ______

☆ **Tell how you solved the problem.**

 Name ____________________

Add Two-Digit Numbers

Find each sum.

4

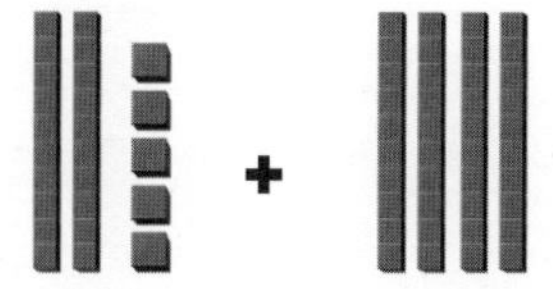

25 + 40

	tens	ones
	2	5
+	4	0

2

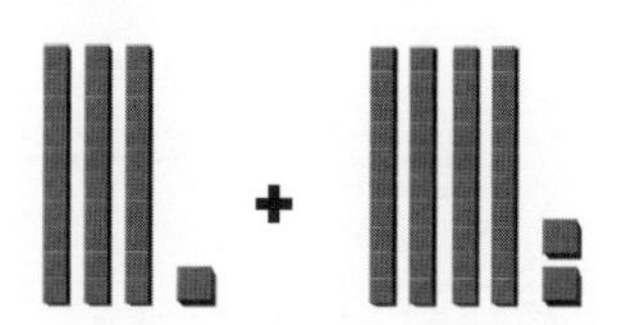

31 + 42

	tens	ones
	3	1
+	4	2

3

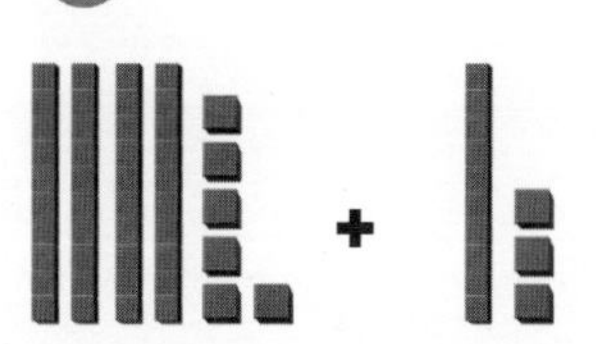

46 + 13

	tens	ones
+		

4

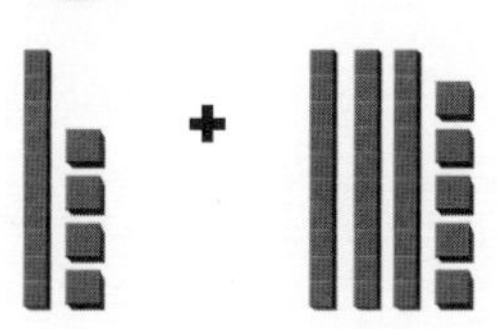

14 + 35

	tens	ones
+		

☆ **Tell how you add.**

 Name ________________________________

Add Two-Digit Numbers

Find each sum.

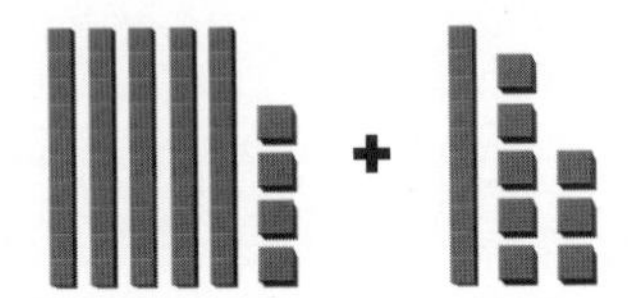

54 + 18

	tens	ones
	5	4
+	1	8

2

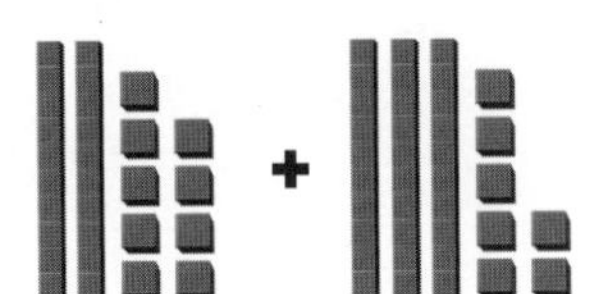

29 + 37

	tens	ones
	2	9
+	3	7

3

36 + 26

	tens	ones
+		

4

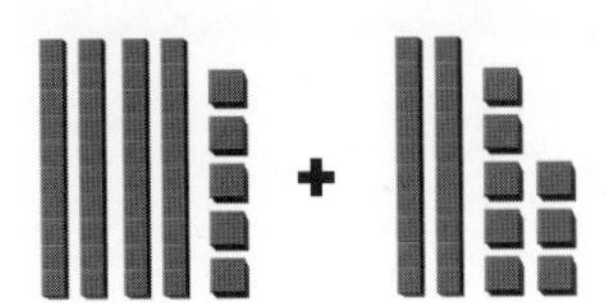

45 + 28

	tens	ones
+		

☆ **Tell how you know when you have to regroup.**

Name ______________________________

Add Two-Digit Numbers

Add two addends. Then add the third addend to the sum.

1 55 + 15 + 16

	tens	ones
	5	5
+	1	5

	tens	ones
+	1	6

2 12 + 48 + 33

	tens	ones
	1	2
+	4	8

	tens	ones
+	3	3

3 54 + 16 + 24

	tens	ones
	5	4
+	1	6

	tens	ones
+	2	4

4 41 + 13 + 27

	tens	ones
	4	1
+	1	3

	tens	ones
+	2	7

 **Tell how you found the sum.**

Assessment

Find each sum.

1

22 + 34

20 + 30 = ______ ⇨ ______

2 + 4 = ______ ⇨ ______

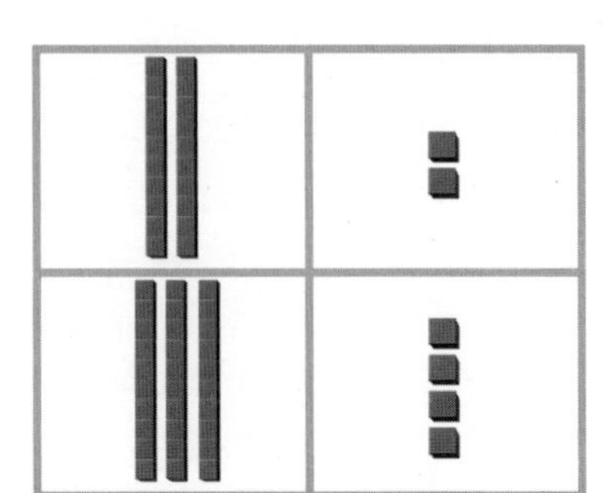

2

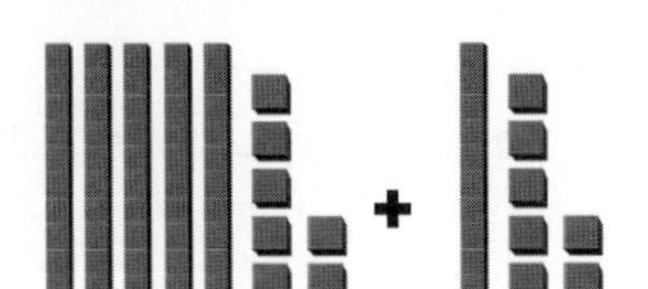

57 + 17

	tens	ones
	5	7
+	1	7

3

33 + 46

	tens	ones
+		

4 **45 + 19 + 31**

	tens	ones
	4	5
+	1	9

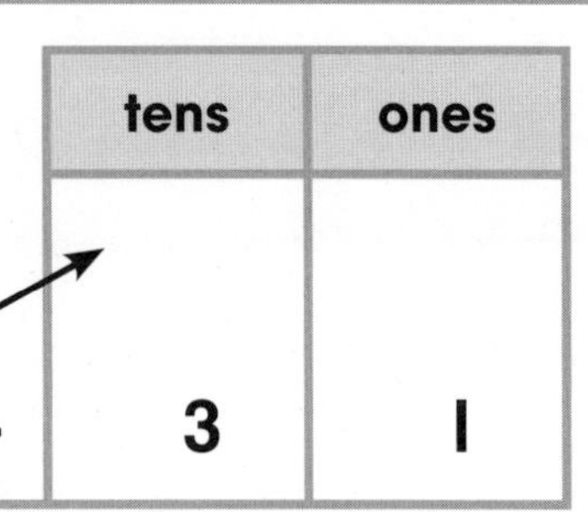

	tens	ones
+	3	1

☆ **Tell how you solved the problem.**

Overview One Hundred More, One Hundred Less

Directions and Sample Answers for Activity Pages

Day 1	See "Model the Skill" below.
Day 2	Read the directions aloud. Remind students that when they subtract, they take away, and that is why they will cross out to subtract. Make sure that students cross out a hundred and not a ten. (Answers: **1.** 151; **2.** 266; **3.** 320; **4.** 78)
Day 3	Read the directions aloud. Tell students that they need to look carefully at the operation symbol to see whether they are adding one hundred or subtracting one hundred. Guide students to see that the number in the hundreds place is either one more or one less, while the numbers in the tens and ones places did not change. (Answers: **1.** 853; **2.** 774; **3.** 588; **4.** 642)
Day 4	Read the directions aloud. Provide each student with a coin. Point out the key that indicates if students toss heads, they should add 100, and if they toss tails, they should subtract 100. Guide students to complete the number sentence by writing the operation symbol in the circle, 100 on the line after it, and the answer after the equals sign. (Answers will vary.)
Day 5	Read the directions aloud. Observe as students complete the page. Do students perform the correct operation? Do they change only the digit in the hundreds place? Use your observations to plan further instruction and review. (Answers: **1.** 115; **2.** 224; **3.** 668; **4.** 877)

Model the Skill

- Hand out the Day 1 activity page and base-ten blocks. Have students show 71 with their blocks. **Say:** *Add a ten. Now how many do you have?* (81) Have students return their blocks to showing 71. **Say:** *Now take away a ten. What number do the blocks show?* (61) Continue to practice adding and subtracting 10 from a given number.
- **Say:** *Today you will be adding one hundred to some numbers.* Have students look at problem 1 and have their blocks show 71. **Ask:** *What number are you adding to 71?* (100) Observe as students add a hundred to their blocks. **Ask:** *What number do your blocks show now?* (171) Have students write the sum. Then guide students to complete problem 2. (156)
- **Say:** *What is the first addend in problem 3?* (139) Observe as students show 1 hundred, 3 tens, and 9 ones blocks. **Ask:** *What are you adding to 139?* (100) *How do you show that?* (Possible answer: I place another hundred block.) *What is the sum of 139 and 100?* (239)

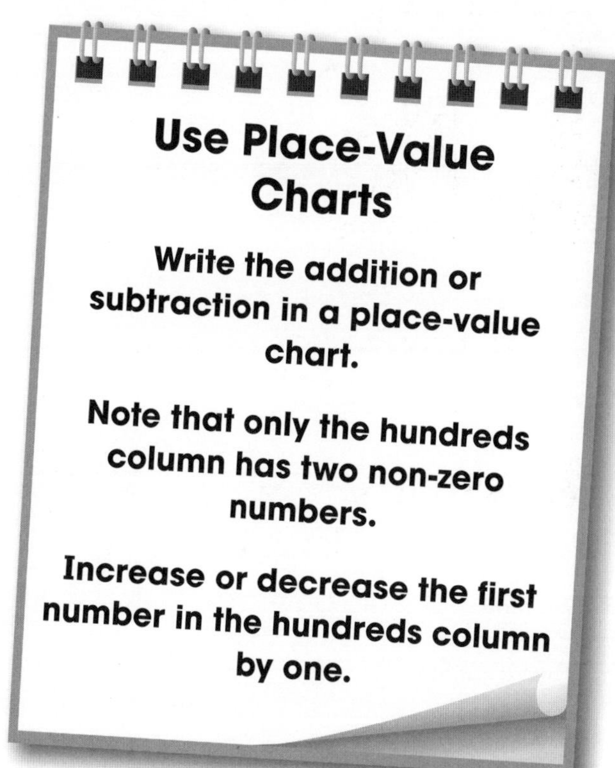

- Help students complete problem 4 by showing 262, adding one hundred, and writing the sum. (362)

One Hundred More, One Hundred Less

Show the addition with blocks. Write the sum.

71 + 100 = ______

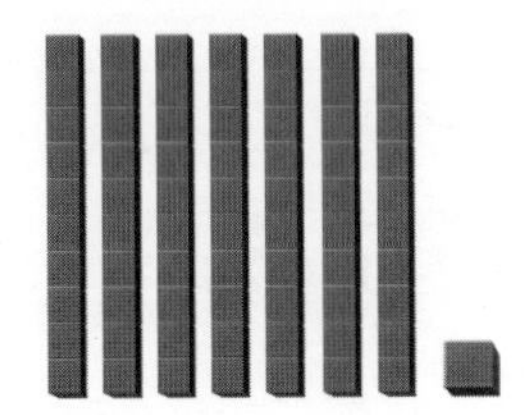

+

2

56 + 100 = ______

+

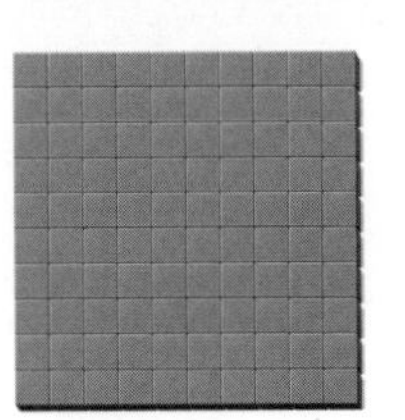

3

139 + 100 = ______

+

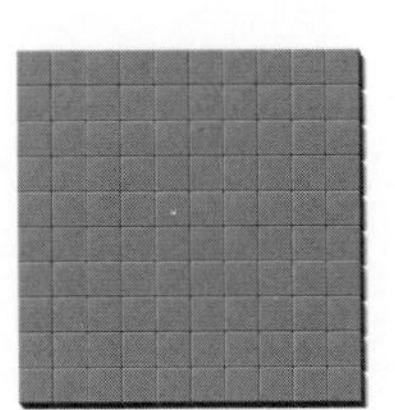

4

262 + 100 = ______

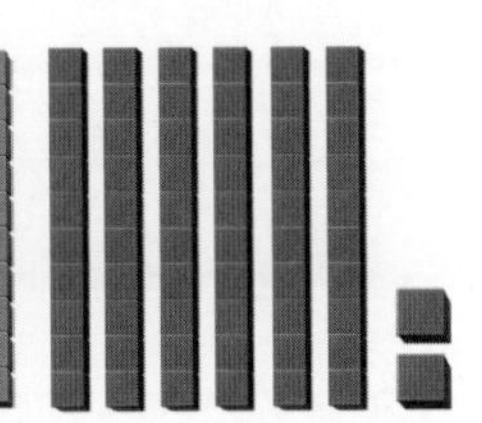

+

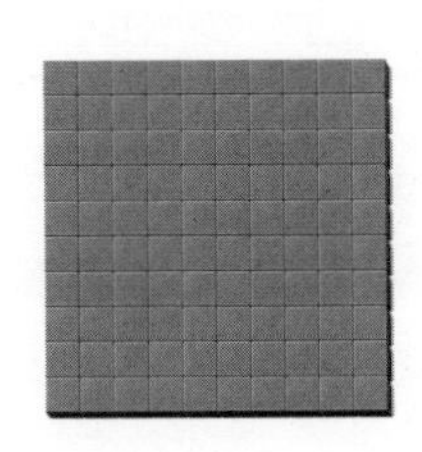

☆ **Tell which place changes when you add one hundred.**

One Hundred More, One Hundred Less

Cross out to subtract. Write the difference.

251 – 100 = ______

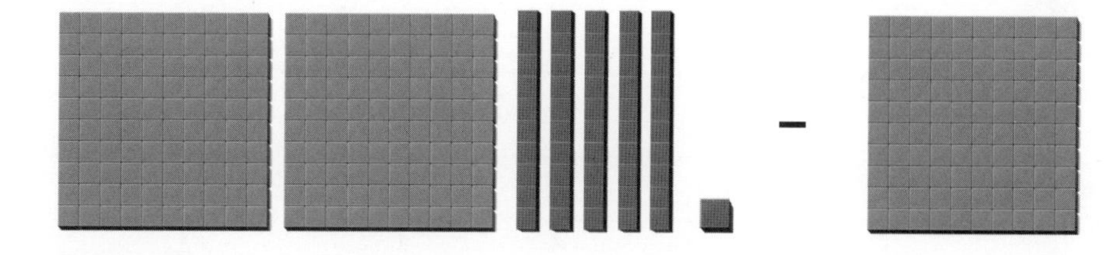

2

366 – 100 = ______

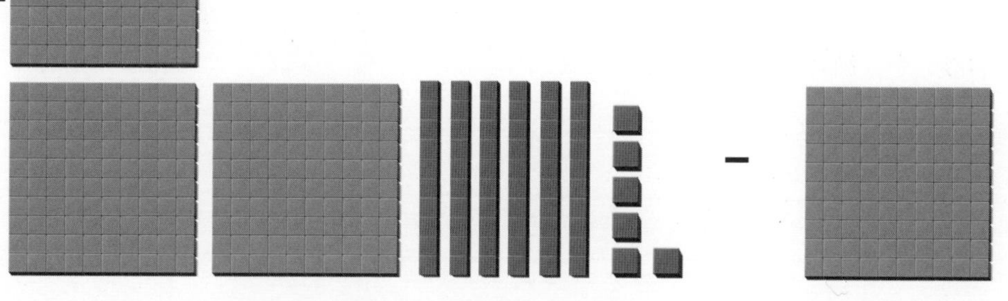

3

420 – 100 = ______

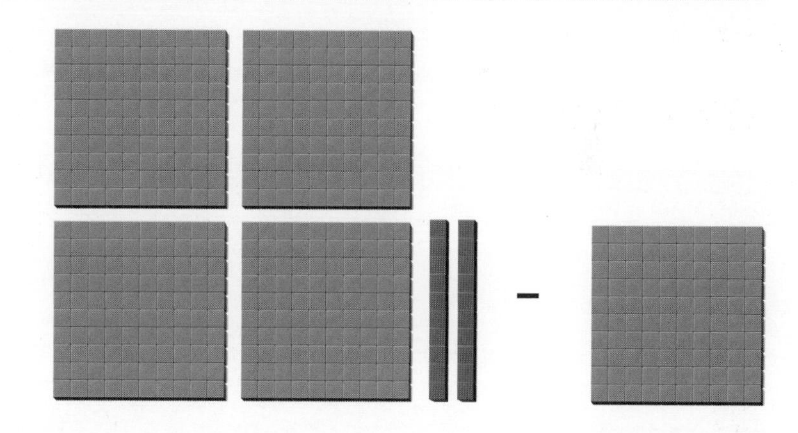

4

178 – 100 = ______

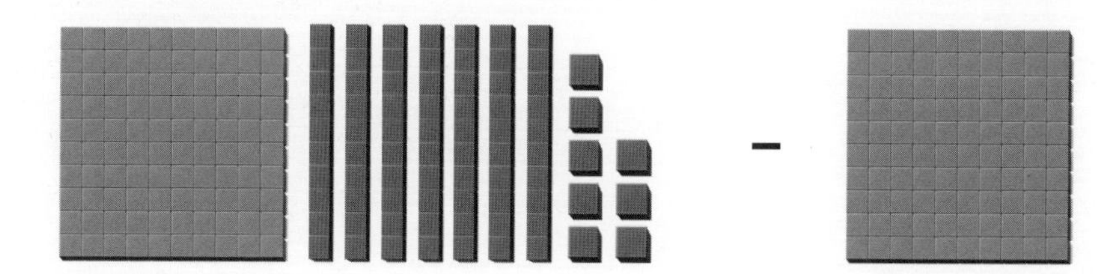

 **Tell how you subtract one hundred.**

One Hundred More, One Hundred Less

Find each sum or difference.

1 753 + 100

	hundreds	tens	ones
	7	5	3
+	1	0	0

2 874 – 100

	hundreds	tens	ones
	8	7	4
–	1	0	0

3 688 – 100

	hundreds	tens	ones
	6	8	8
–	1	0	0

4 542 + 100

	hundreds	tens	ones
	5	4	2
+	1	0	0

☆ **Tell how a place-value chart helps you solve the problem.**

One Hundred More, One Hundred Less

Flip a coin. Follow the key to add or subtract. Complete the number sentence.

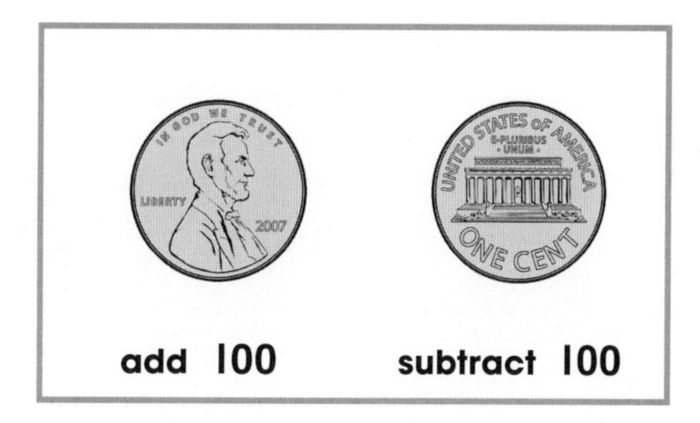

1. 447 ◯ ______ = ______

2. 865 ◯ ______ = ______

3. 391 ◯ ______ = ______

4. 607 ◯ ______ = ______

☆ **Tell how the first number compares with the answer.**

Assessment

Solve each problem.

215 − 100 = ______  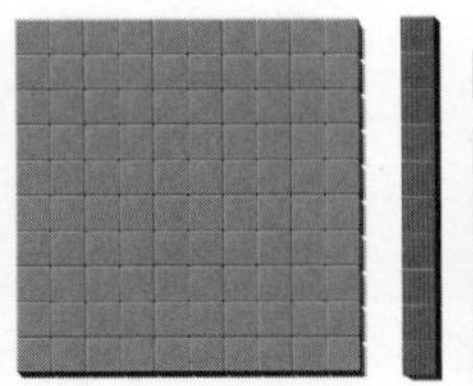 - 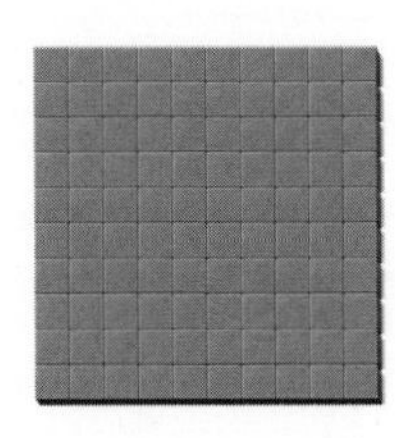

2

124 + 100 = ______ 

---

3 568 + 100

	hundreds	tens	ones
	5	6	8
+	1	0	0

4

977 − 100 = __________

☆ **Tell how you know solved the problem.**

Overview Add Three-Digit Numbers

Directions and Sample Answers for Activity Pages

Day 1	See "Model the Skill" below.
Day 2	Read the directions aloud. Tell students that the blocks and the place-value charts are shown to help them add each place—starting with the ones and working left to the hundreds. Point out that students are to write the addends in the place-value chart in problems 3 and 4. Observe that students add the ones, then the tens, and finally the hundreds. (Answers: **1.** 284; **2.** 366; **3.** 457; **4.** 329)
Day 3	Read the directions aloud. Tell students that the problems on this page will require them to regroup. Remind students that when they regroup 10 ones as 1 ten, they write the ones digit at the bottom of the column and the regrouped digit at the top of the tens column. Tell students that they use a similar procedure when regrouping 10 tens as 1 hundred. Check that when students regroup 10 tens as 1 hundred, they write the regrouped digit at the top of the hundreds column and add it when adding the hundreds. (Answers: **1.** 282; **2.** 384; **3.** 374; **4.** 417)
Day 4	Read the directions aloud. Tell students that they will add the digits without the help of base-ten blocks on this page. Remind students to add the ones column first, the tens column second, and the hundreds column last. Check that students are regrouping 10 ones as 1 ten when there are more than 10 ones, and regrouping 10 tens as 1 hundred when there are more than 10 tens. (Answers: **1.** 784; **2.** 898; **3.** 637; **4.** 536)
Day 5	Read the directions aloud. Observe as students complete the page. Do students add the ones, then the tens, and then the hundreds? Do they remember to add the regrouped digit? Use your observations to plan further instruction and review. (Answers: **1.** 447; **2.** 321; **3.** 763; **4.** 891)

Model the Skill

- Hand out the Day 1 activity page and base-ten blocks. **Say:** *Today we are going to add 2 three-digit numbers. A three-digit number has hundreds, tens, and ones.* Show the two addends in problem 1 with your blocks. Observe as students build the numbers. **Say:** *Add the hundreds together. How many are there?* (2) *That means that 100 plus 100 is 200.* Guide students to write that sum on the activity page. **Ask:** *What amounts are we going to add next?* (20 and 30) Guide students to join the tens and write 5 tens as 50. Then guide students to join the ones and write the sum on the page. (9) **Say:** *Now add the three place-value sums. What is the total sum?* (259)

- Have students show the addends in problem 2. **Ask:** *What is the first step?* (Add the hundreds.) *What is the sum?* (300) *What is the second step?* (Add the tens.) *What is the sum?* (30) *What is the third step?* (Add the ones.) *What is the sum?* (11) Have students add the two numbers in the tens place to find the total. (341)

- Help students complete the activity page by building the addends with their blocks, adding each place, writing the partial sums, and then adding the partial sums. Guide students to see that they have to add in some places when finding the final sum. (Answers: **3.** 300 + 150 + 7 = 457; **4.** 200 + 120 + 11 = 331)

Add Three-Digit Numbers

Show each addend with blocks. Add each place. Add the sums.

1

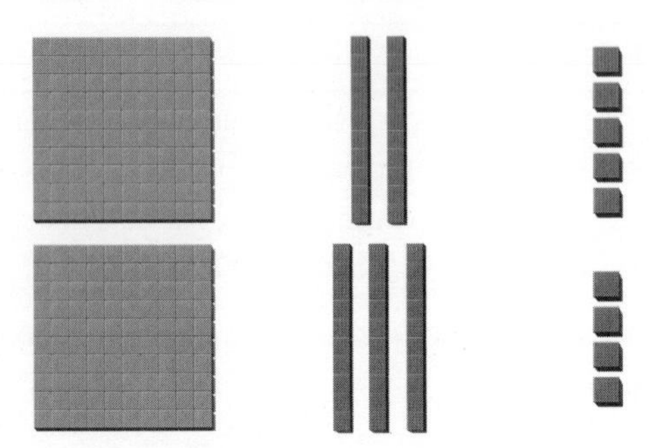

125 + 134

100 + 100 = ______ ➩ ______

20 + 30 = ______ ➩ ______

5 + 4 = ______ ➩ ______

\+ ______

2

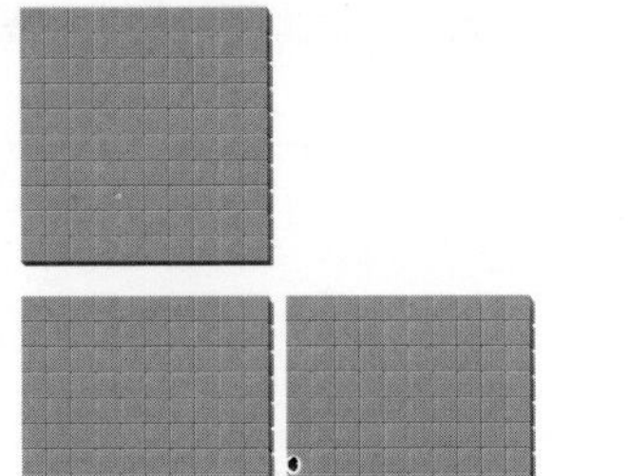

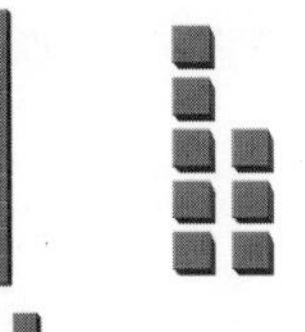

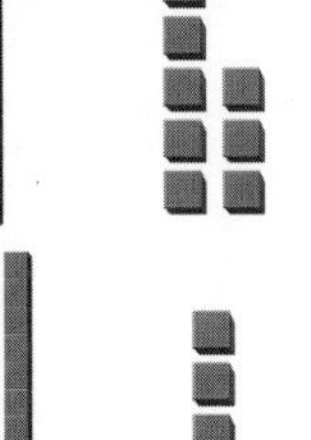

118 + 223

100 + 200 = ______ ➩ ______

10 + 20 = ______ ➩ ______

8 + 3 = ______ ➩ ______

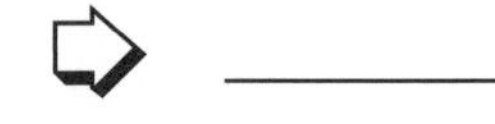

\+ ______

3

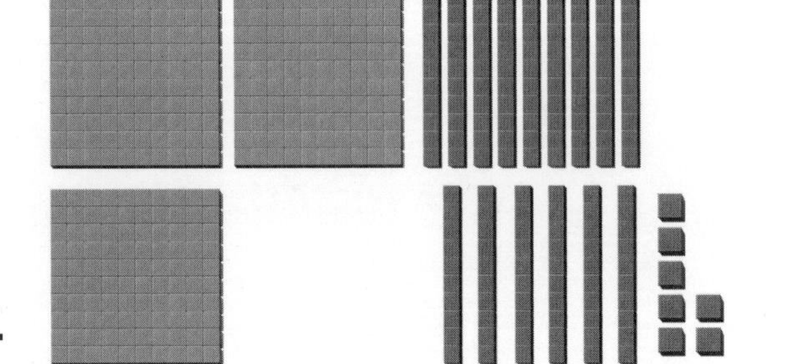

290 + 167

200 + 100 = ______ ➩ ______

90 + 60 = ______ ➩ ______

0 + 7 = ______ ➩ ______

\+ ______

4

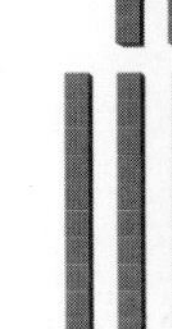

159 + 172

100 + 100 = ______ ➩ ______

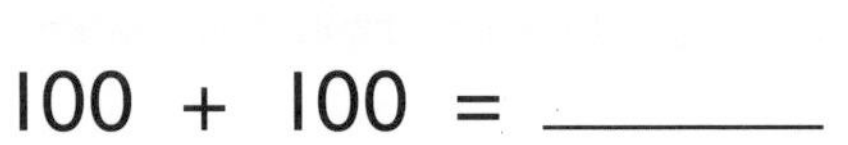

50 + 70 = ______ ➩ ______

9 + 2 = ______ ➩ ______

\+ ______

☆ **Tell how you found the sum.**

Name ______________________________

Add Three-Digit Numbers

Find each sum.

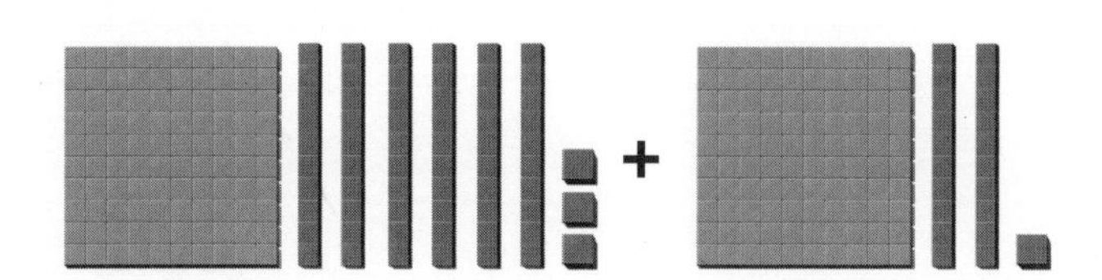

163 + 121

	hundreds	tens	ones
	1	6	3
+	1	2	1

2

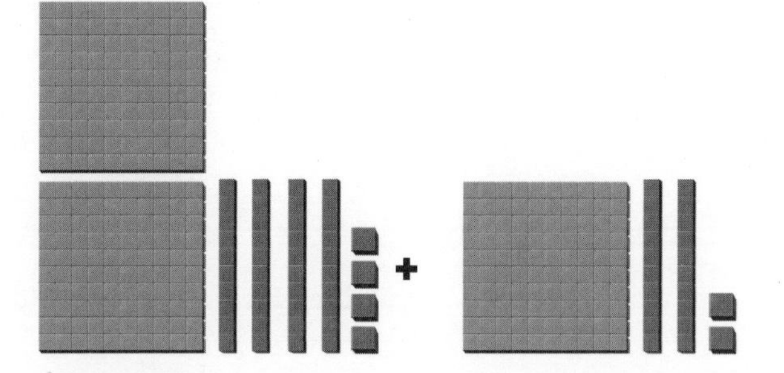

244 + 122

	hundreds	tens	ones
	2	4	4
+	1	2	2

3

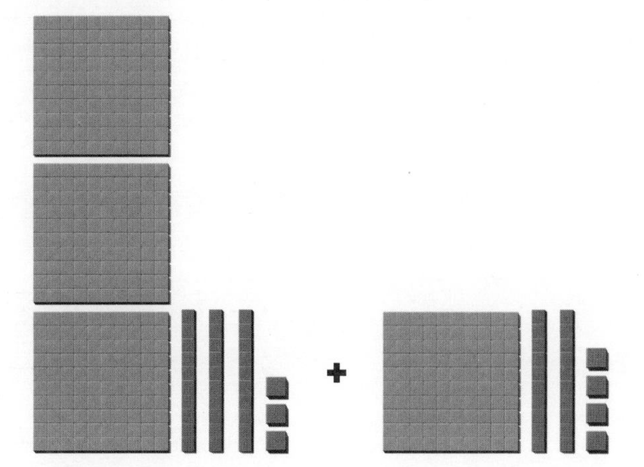

333 + 124

	hundreds	tens	ones
+			

4

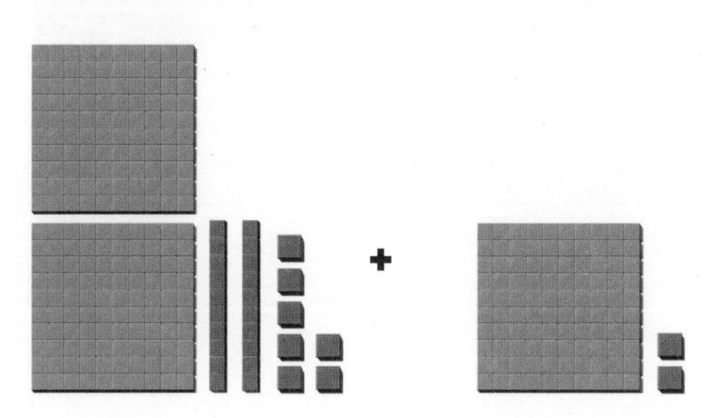

227 + 102

	hundreds	tens	ones
+			

☆ **Tell how you add three-digit numbers.**

Add Three-Digit Numbers

Find each sum.

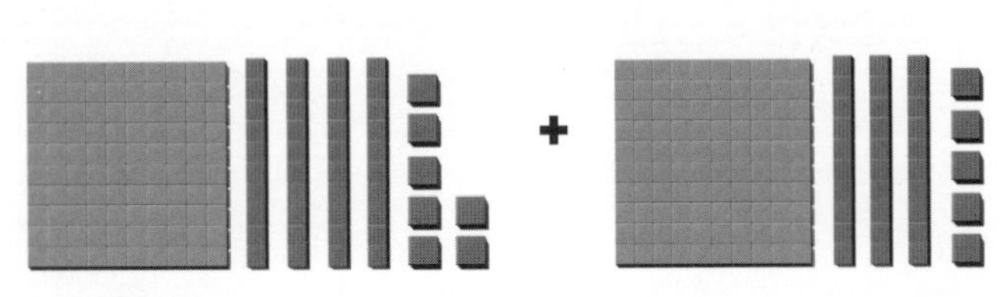

147 + 135

	hundreds	tens	ones
	1	4	7
+	1	3	5

2

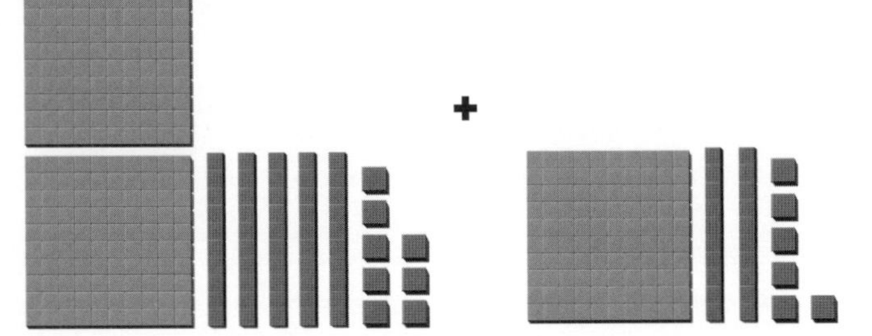

258 + 126

	hundreds	tens	ones
	2	5	8
+	1	2	6

3

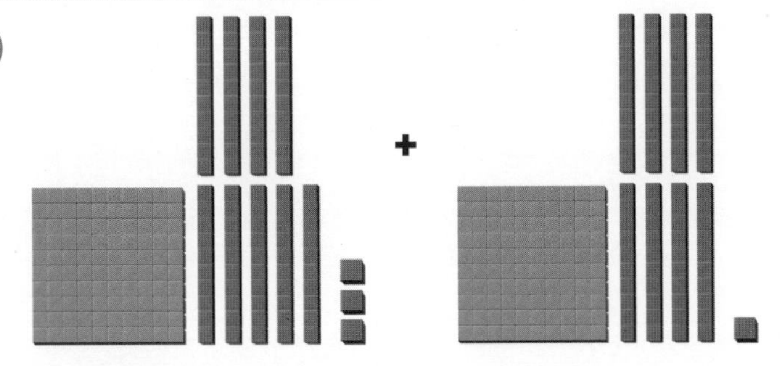

193 + 181

	hundreds	tens	ones
	1	9	3
+	1	8	1

4

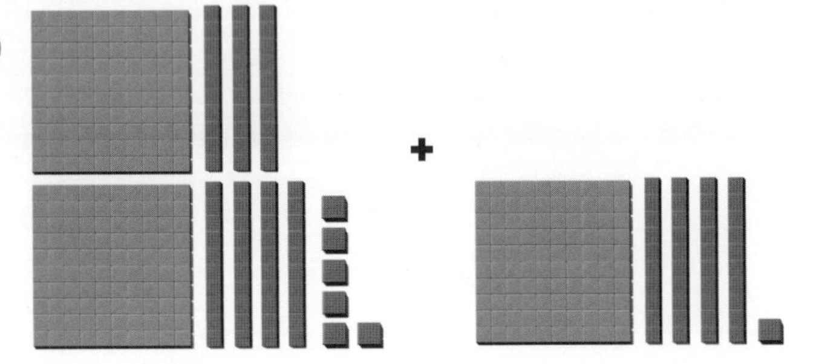

276 + 141

	hundreds	tens	ones
	2	7	6
+	1	4	1

 **Tell how you solved the problem.**

Add Three-Digit Numbers

Find each sum.

1 652 + 132

	hundreds	tens	ones
	6	5	2
+	1	3	2

2 569 + 329

	hundreds	tens	ones
	5	6	9
+	3	2	9

3 473 + 164

	hundreds	tens	ones
	4	7	3
+	1	6	4

4 287 + 249

	hundreds	tens	ones
	2	8	7
+	2	4	9

☆ **Tell how you know when you have to regroup.**

 Name ______________________

Assessment

Find each sum.

1. 194 + 253

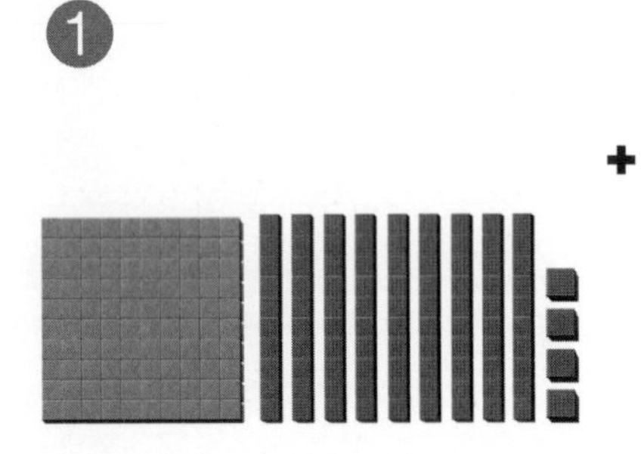
+
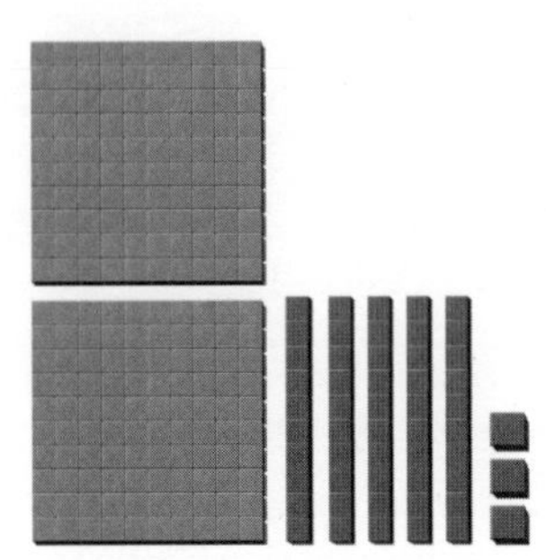

	hundreds	tens	ones
	1	9	4
+	2	5	3

2. 179 + 142

	hundreds	tens	ones
	1	7	9
+	1	4	2

3. 343 + 420

	hundreds	tens	ones
	3	4	3
+	4	2	0

4. 685 + 206

	hundreds	tens	ones
	6	8	5
+	2	0	6

☆ **Tell how you solved the problem.**

Overview Use Strategies to Subtract

Directions and Sample Answers for Activity Pages

Day 1	See "Model the Skill" below.
Day 2	Read the directions aloud. Remind students that addition and subtraction are related. Explain that they can think of a related addition fact to help them subtract. Point out that the think bubbles tell them which related fact to use. Guide students to see that the missing addend in the addition fact is the difference of the subtraction. (Answers: **1.** 8; **2.** 6; **3.** 8; **4.** 7)
Day 3	Read the directions aloud. Tell students that a place-value chart can be used to help them focus on subtracting the numerals in each place. Point out that the equations are written in a place-value chart. Remind students to subtract the ones before subtracting the tens. (Answers: **1.** 34; **2.** 22; **3.** 31; **4.** 42)
Day 4	Read the directions aloud. Tell students that another strategy is to use models, such as base-ten blocks. Point out that regrouping is needed in the problems on this page. Guide students to cross out a ten and draw 10 ones to regroup. Remind students to cross out the ones to subtract, and then the tens. Spend as much time as needed to help students see that they need to decompose 1 ten into 10 ones. (Answers: **1.** 32; **2.** 44; **3.** 28; **4.** 15)
Day 5	Read the directions aloud. Observe as students complete the page. Do students use different strategies in each problem? Do they remember to regroup when needed? Do they subtract the ones before the tens? Use your observations to plan further instruction and review. (Answers: **1.** 69; **2.** 7; **3.** 32; **4.** 23)

Model the Skill

- Hand out the Day 1 activity page.
- **Say:** *You can use different strategies to subtract. You can count back by 1, 2, or 3 on a number line.* Have students look at problem 1 and circle the first number in the equation on the number line. **Say:** *You can draw two jumps to count back 2 from 28.* Show students how to draw a curved line from 28 to 27 and then to 26. Encourage students to count back aloud as they draw the jumps. **Ask:** *On what number did you land?* (26)
- Have students look at problem 2. **Ask:** *What number will you circle on the number line?* (42) *How many jumps will you make?* (3) Observe as students draw the jumps and count back aloud. **Ask:** *What is 42 minus 3?* (39)
- Help students complete the activity page by counting back on the number line. Remind students to circle the beginning number, draw jumps to the left to show the number they are counting back, and to write the number on which they land as the difference. (Answers: **3.** 60; **4.** 36)

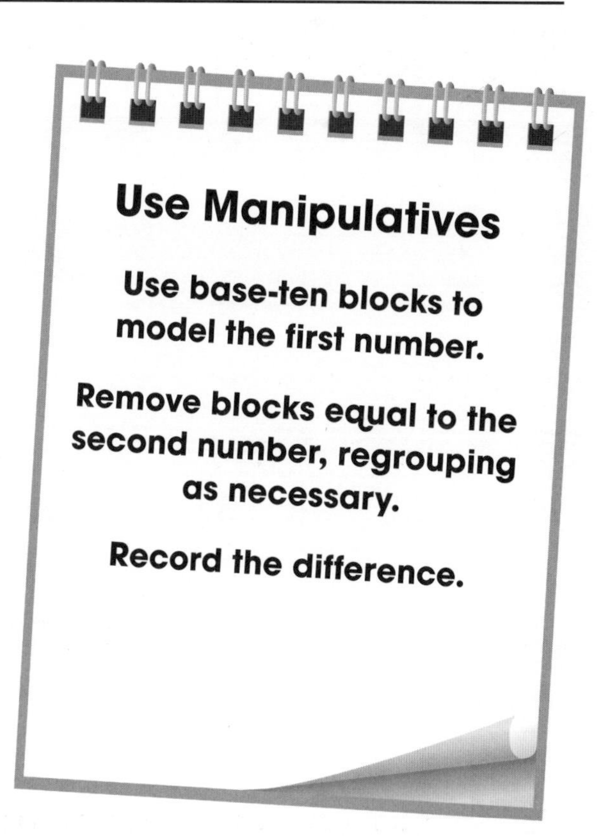

 Name ______________________

Use Strategies to Subtract

Count back to subtract. Write each difference.

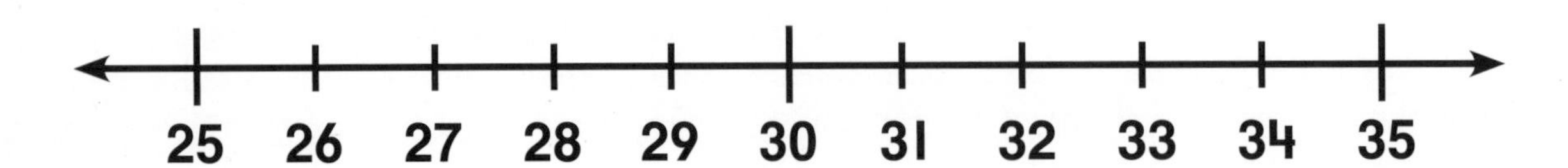

28 − 2 = ______

2

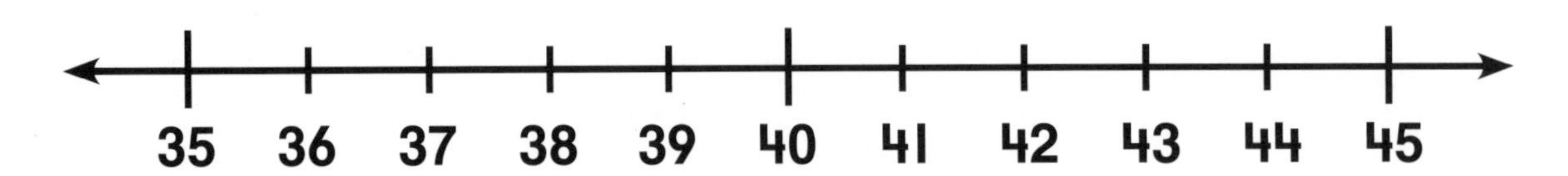

42 − 3 = ______

3

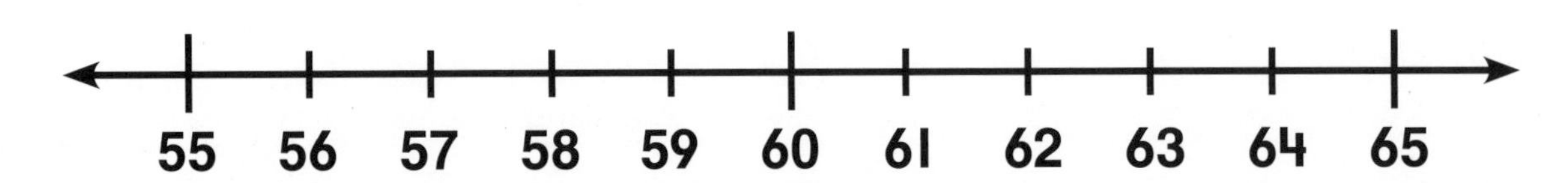

63 − 3 = ______

4

35 36 37 38 39 40 41 42 43 44 45

37 − 1 = ______

☆ **Tell how you can use a number line to subtract.**

Name ______________________________

Use Strategies to Subtract

Use a related addition fact to subtract.

1. $17 - 9 =$ ______

Think:
$9 + ? = 17$

2. $14 - 8 =$ ______

Think:
$8 + ? = 14$

3. $15 - 7 =$ ______

Think:
$7 + ? = 15$

4. $11 - 4 =$ ______

Think:
$4 + ? = 11$

☆ **Tell how you can use an addition fact to help you subtract.**

 Name ____________________

Use Strategies to Subtract

Find each difference.

1. 38 − 4

	tens	ones
	3	8
−		4

2. 29 − 7

	tens	ones
	2	9
−		7

3. 37 − 6

	tens	ones
	3	7
−		6

4. 46 − 4

	tens	ones
	4	6
−		4

☆ **Tell how you used place value to subtract.**

Name ____________________

Use Strategies to Subtract

Find each difference.

1. 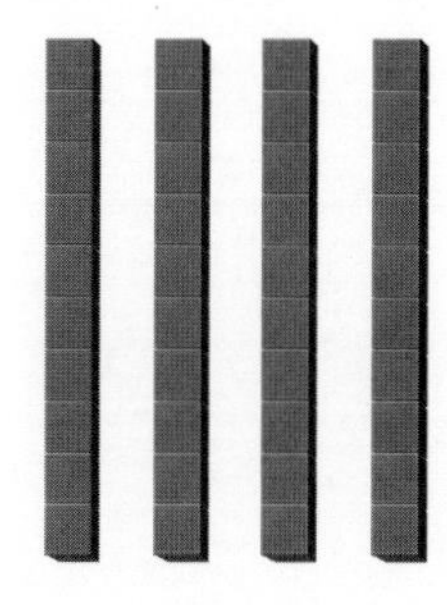

40 − 8

	tens	ones
	4	0
−		8

2. 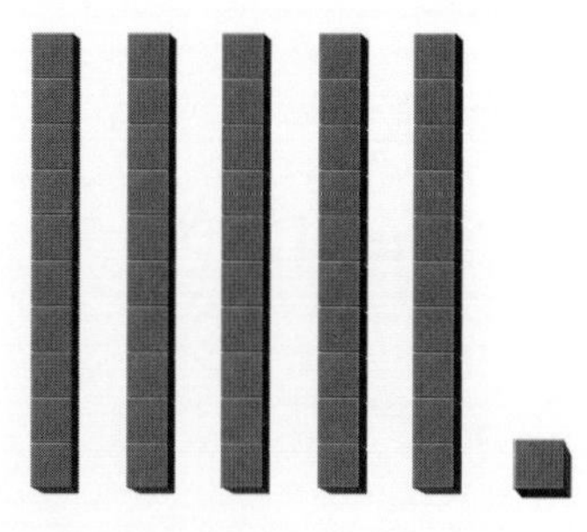

51 − 7

	tens	ones
	5	1
−		7

3. 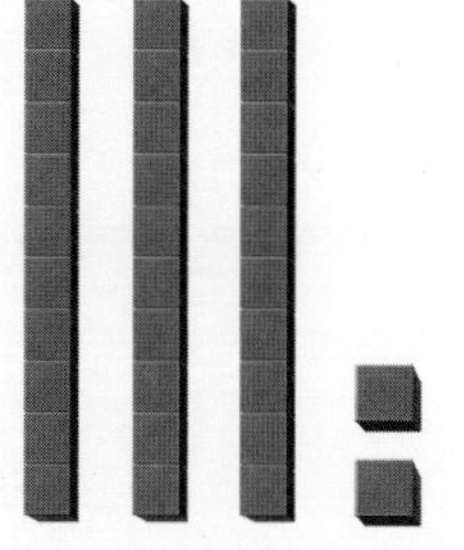

32 − 4

	tens	ones
	3	2
−		4

4. 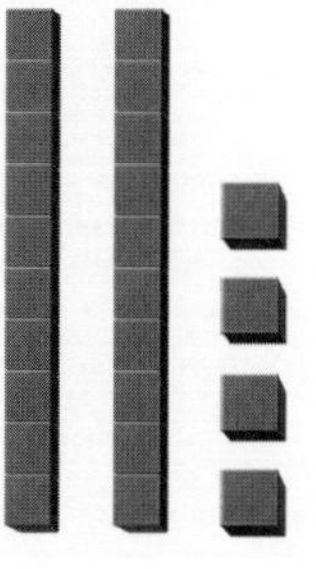

24 − 9

	tens	ones
	2	4
−		9

☆ **Tell how you used blocks to subtract.**

 Name ______________________

Assessment

Find each difference.

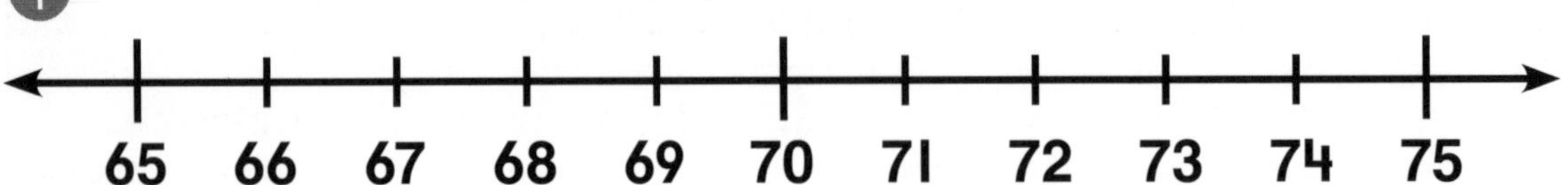

71 − 2 = ______

2

12 − 5 = ______

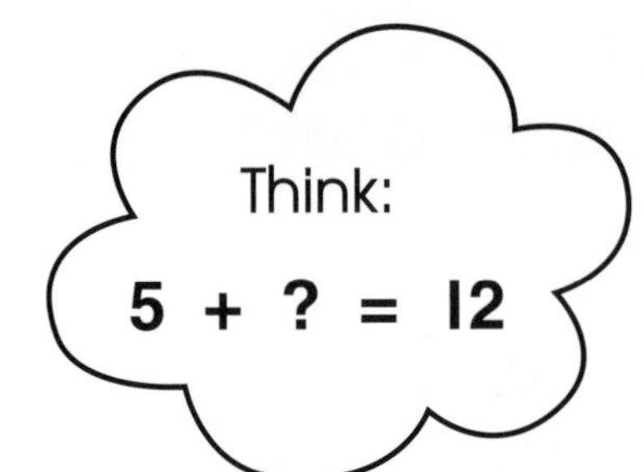

3 39 − 7

	tens	ones
	3	9
−		7

4

31 − 8

	tens	ones
	3	1
−		8

☆ **Tell how you solved the problem.**

Overview Subtract Two-Digit Numbers

Directions and Sample Answers for Activity Pages

Day 1	See "Model the Skill" below.
Day 2	Read the directions aloud. Tell students that a place-value chart can be used to show the subtraction for each place. Point out that the subtraction is written in the place-value chart in problems 1 and 2 but that they will need to write in the chart for problems 3 and 4. Remind students to subtract the ones before subtracting the tens. Guide students to cross out the pictured blocks to see the subtraction. (Answers: **1.** 11; **2.** 23; **3.** 39; **4.** 26)
Day 3	Read the directions aloud. Explain that regrouping is needed in the problems because there are not enough ones to subtract. Have students cross out a ten and draw 10 ones to show the trade. Then guide students to cross out the numbers in the vertical subtraction and write the regrouped values. Spend the time needed to help students see the connection between the visual and symbolic regroupings. (Answers: **1.** 12; **2.** 17; **3.** 6; **4.** 19)
Day 4	Read the directions aloud. Tell students to subtract their way along the path. Allow them to use base-ten blocks or place-value charts if needed. Students who subtract correctly should have a final difference of zero. Tell students that since addition and subtraction are related, they can add to check their subtraction. (Answers: **1.** 62; **2.** 44; **3.** 19; **4.** 0)
Day 5	Read the directions aloud. Have students complete the page. Do students subtract the ones before the tens? Do they regroup when needed? Do they show the regrouping correctly? Use your observations to plan further instruction. (Answers: **1.** 23; **2.** 20; **3.** 8; **4.** 47)

Model the Skill

- Hand out the Day 1 activity page and base-ten blocks. **Say:** *You can use base-ten blocks to model subtraction.* Have students look at problem 1 and note the subtraction sentence. **Say:** *The beginning number is 26. How do you show 26 with your blocks?* (2 tens, 6 ones) *The number sentence says that we need to subtract 14. How do you show that with the blocks?* (Take away 1 ten and 4 ones.) Guide students to model the subtraction. **Ask:** *What is left?* (12) Observe as students write the difference. You may wish to have students cross out the blocks pictured on the activity page to show the subtraction.
- **Ask:** *What are you going to show with your blocks to model the subtraction in problem 2?* (Possible answer: Show 3 tens and 2 ones. Then take away 1 ten and 1 one.) Observe as students model the subtraction. **Ask:** *What is the difference between 32 and 11?* (21)

Use Manipulatives

Use base-ten blocks to model the subtraction.

Take away blocks to show the subtraction, regrouping as necessary.

Record the subtraction in both horizontal (equation) and vertical formats.

- Have students model the beginning number in problem 3. **Ask:** *Can you take away 1 ten and 7 ones from 3 tens?* (no) Show students how to trade 1 ten for 10 ones to show 30 in a different way. **Say:** *Now you have some ones to take away. What is left?* (1 ten, 3 ones; 13)
- Help students complete problem 4 by trading 1 ten for 10 ones and then taking away the amount shown. You can have students draw the trade on the activity page. (4)

Subtract Two-Digit Numbers

Model the subtraction with blocks. Write the difference.

1. 

$26 - 14 =$ ______

2. 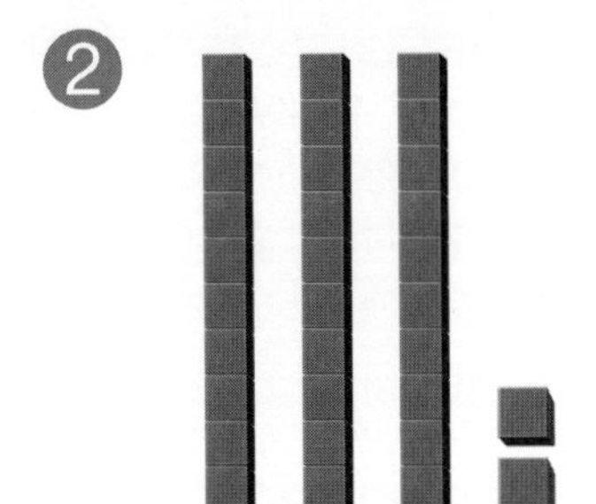

$32 - 11 =$ ______

3. $30 - 17 =$ ______

4. $23 - 19 =$ ______

☆ **Tell how you regroup blocks to subtract.**

Subtract Two-Digit Numbers

Find each difference.

1. 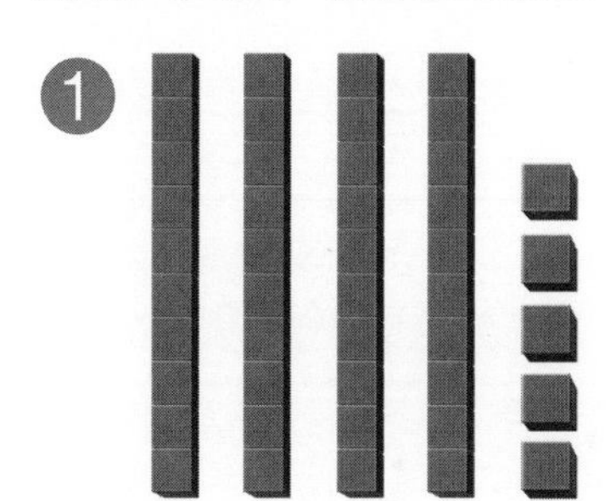

45 − 34

	tens	ones
	4	5
−	3	4

2. 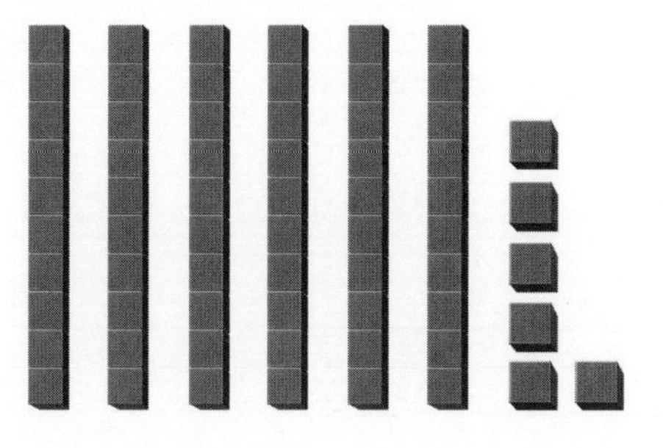

66 − 43

	tens	ones
	6	6
−	4	3

3. 

59 − 20

	tens	ones
−		

4. 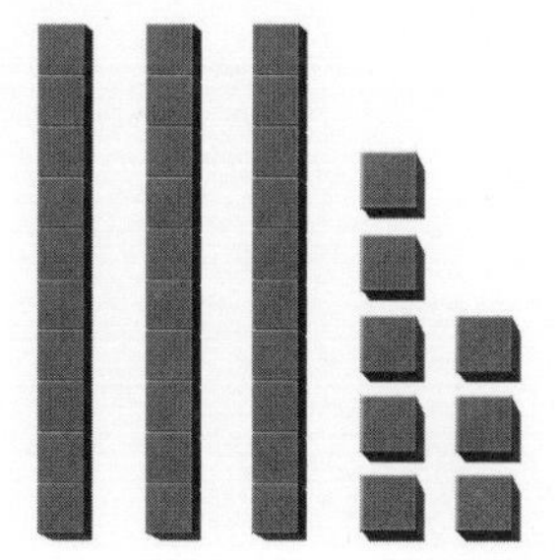

38 − 12

	tens	ones
−		

☆ **Tell how you subtract.**

Name ____________________

Subtract Two-Digit Numbers

Find each difference.

1. 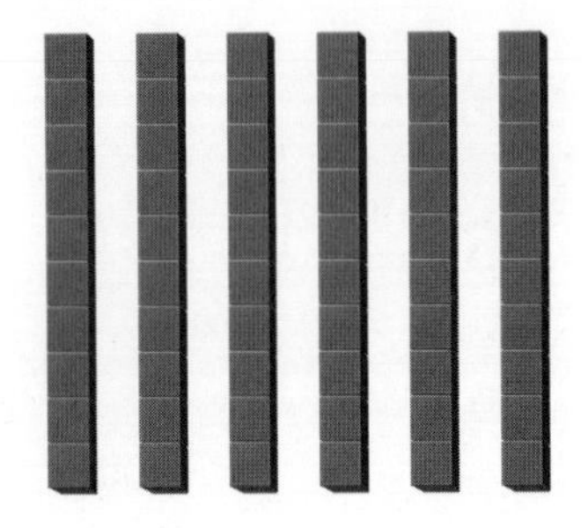

60 − 48

	tens	ones
	6	0
−	4	8

2. 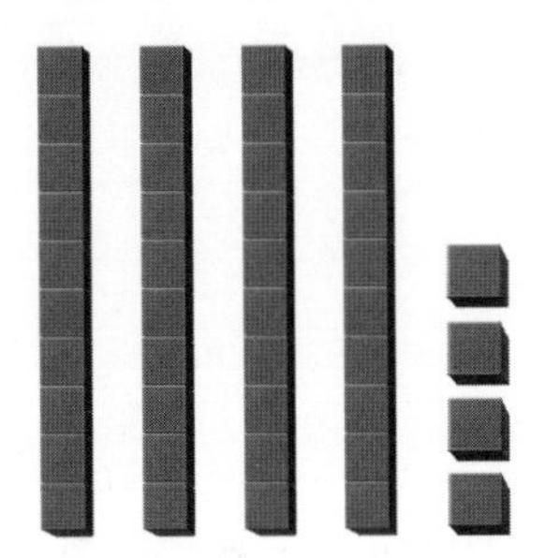

44 − 27

	tens	ones
	4	4
−	2	7

3. 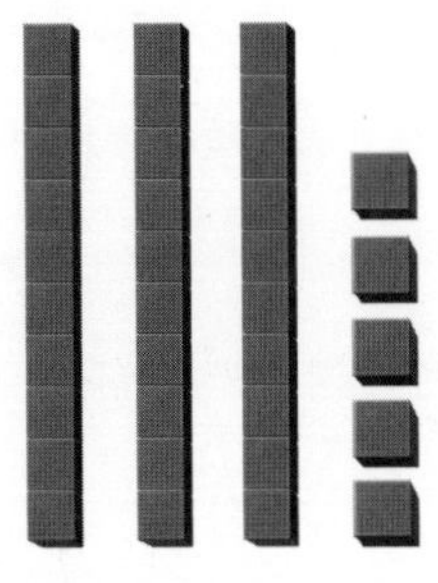

35 − 29

	tens	ones
−		

4. 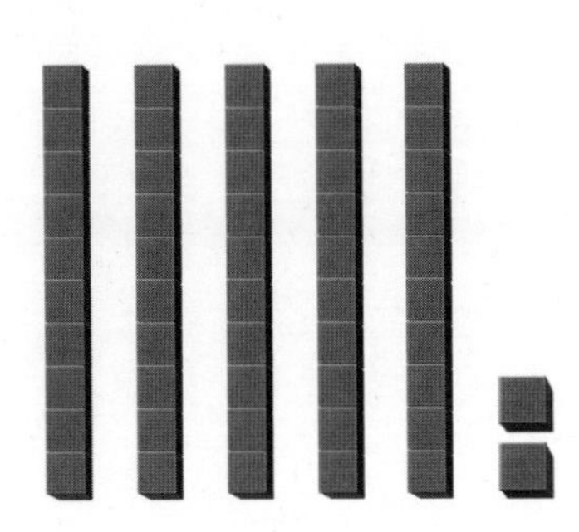

52 − 33

	tens	ones
−		

☆ **Tell how you know when to regroup.**

Subtract Two-Digit Numbers

Follow the path. Subtract each number. Write each difference in a blank space.

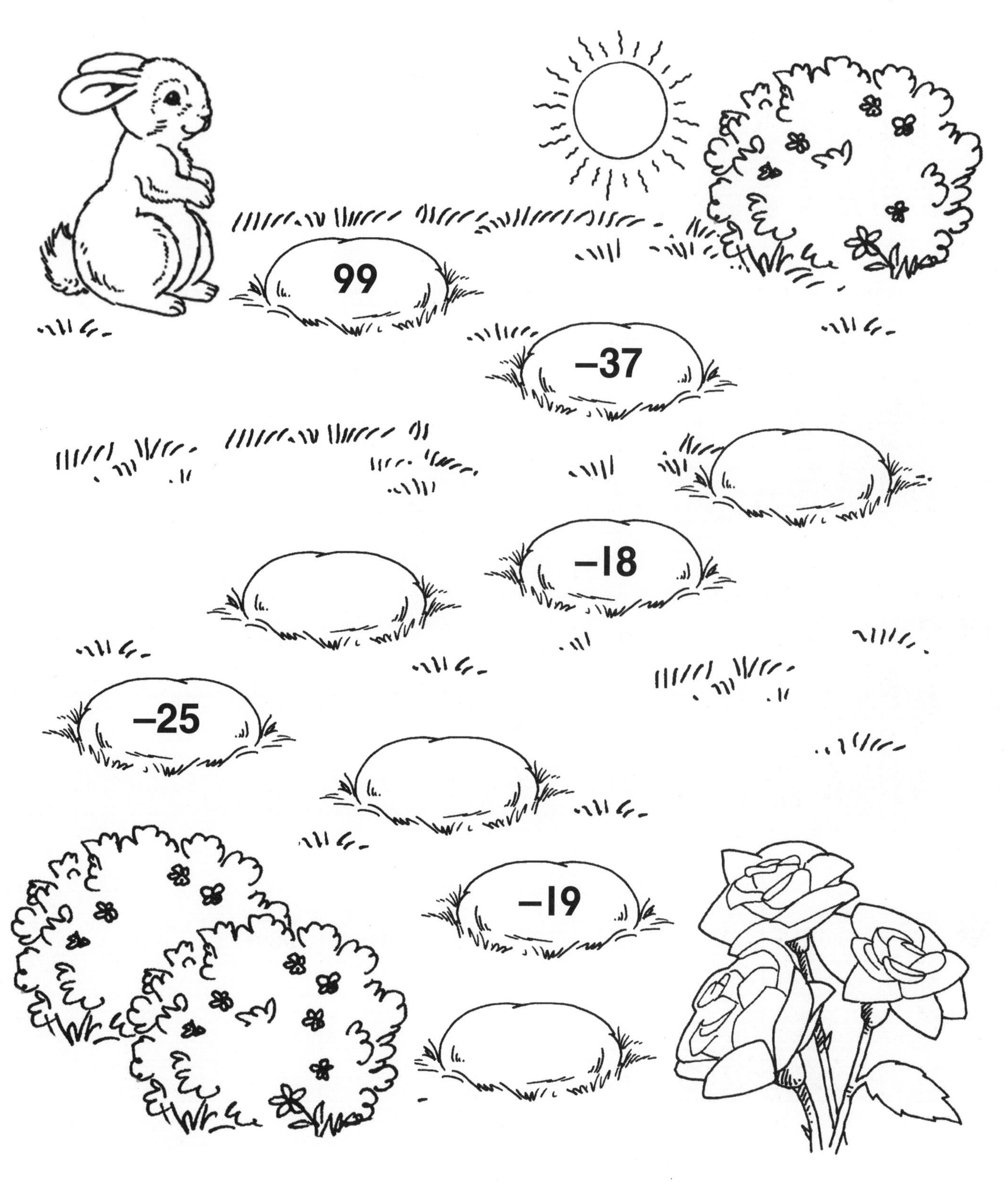

☆ **Tell how you know you subtracted correctly.**

Name ____________________

Assessment

Find each difference.

1. 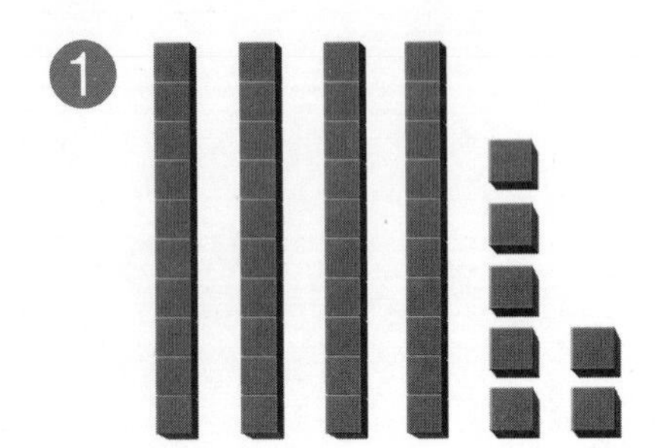

$47 - 24 =$ ______

2. $36 - 16$

	tens	ones
−		

3. $21 - 13$

	tens	ones
−		

4. $62 - 15$

	tens	ones
−		

☆ **Tell how you solved the problem.**

Overview Subtract Three-Digit Numbers

Directions and Sample Answers for Activity Pages

Day 1	See "Model the Skill" below.
Day 2	Read the directions aloud. Tell students that a place-value chart can help them record the subtraction that they show with the model. Remind students to subtract the ones first, then the tens, and then the hundreds. Guide students to cross out the pictured blocks to see the subtraction. (Answers: **1.** 215; **2.** 233; **3.** 224; **4.** 124)
Day 3	Read the directions aloud. Point out that regrouping is needed in the problems on this page. Guide students to cross out a ten and draw 10 ones for problems 1 and 2. Then for problems 3 and 4, they should cross out a hundred and draw 10 tens. Demonstrate how to cross out the numbers in the vertical subtraction and write the regrouped values. Help students see the connection between the visual and the symbolic regroupings. (Answers: **1.** 223; **2.** 233; **3.** 285; **4.** 43)
Day 4	Read the directions aloud. Tell students to use the place-value chart to help them subtract. Explain that the horizontal equations are the same as the subtraction in the charts. Remind students to subtract the ones before the tens, and then the tens before the hundreds. Have students cross out and show the regroupings. (Answers: **1.** 413; **2.** 425; **3.** 365; **4.** 344)
Day 5	Read the directions aloud. Observe as students complete the page. Do students subtract the ones first? Do they regroup when needed? Do they cross out to show the regrouping? Use your observations to plan further instruction and review. (Answers: **1.** 124; **2.** 224; **3.** 325; **4.** 664)

Model the Skill

- Hand out the Day 1 activity page and base-ten blocks. **Say:** *Today, we are going to subtract three-digit numbers. Show the first amount with your blocks.* Have students show the number. **Ask:** *What did you show?* (4 hundreds, 5 tens, and 6 ones) *You need to subtract 214. How do you do that?* (Take away 2 hundreds, 1 ten, and 4 ones.) *What do you have left?* (2 hundreds, 4 tens, 2 ones) Have students write the number on their activity page as the difference between the two numbers. You can have students cross out the pictured blocks to record the subtraction.

- Direct students to model the number in problem 2. **Ask:** *What are you going to take away from 324?* (111; 1 hundred, 1 ten, 1 one) *What is the difference between 324 and 111?* (213)

- Have students show 250 with their blocks for problem 3. **Say:** *We need to subtract 134, but there are not any ones blocks to take away. What do we need to do?* (Possible answer: trade 1 ten for 10 ones) Have students make the trade and show them how the blocks still represent the number 250. **Say:** *Now you can subtract. What do you have left?* (116)

- Help students complete problem 4 by modeling 305, trading 1 hundred for 10 tens, and taking away 272. You may wish to have students draw the trade on the activity page. (33)

Use Manipulatives

Use base-ten blocks to model the first number.

Take away blocks to show the subtraction, regrouping when needed.

Record the subtraction in a place-value chart, showing the regrouping.

 Name ____________________

Subtract Three-Digit Numbers

Model the subtraction with blocks. Write the difference.

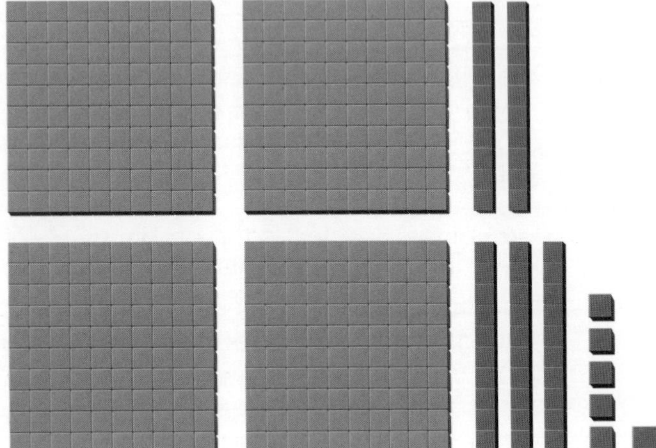

$456 - 214 =$ ______

2

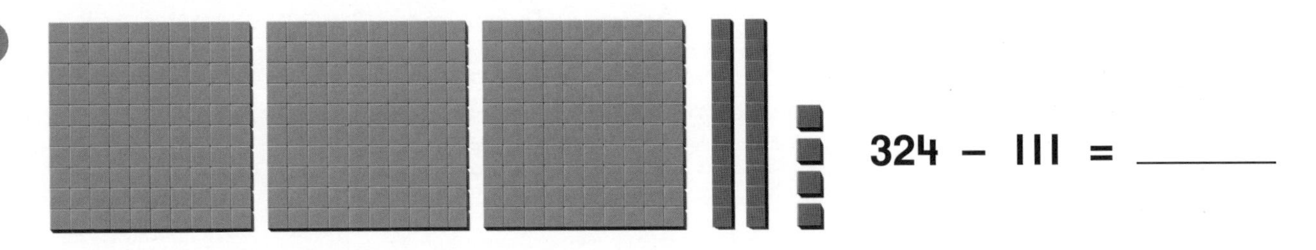

$324 - 111 =$ ______

3

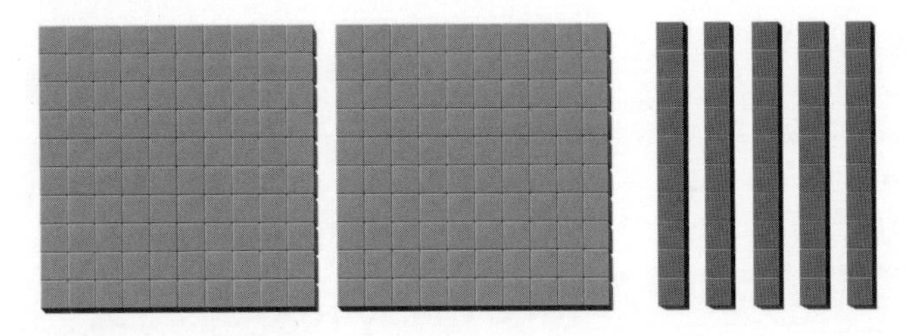

$250 - 134 =$ ______

4

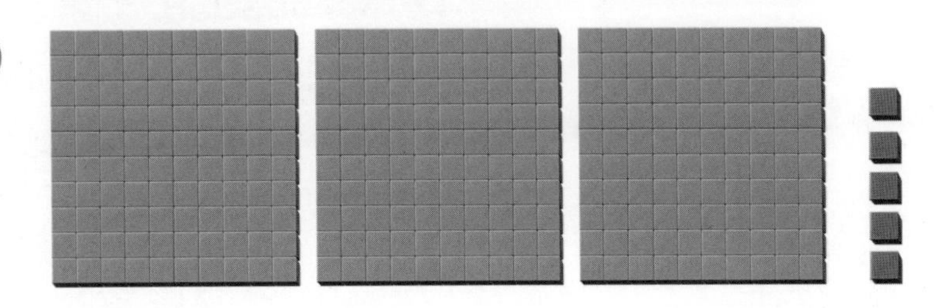

$305 - 272 =$ ______

☆ **Tell how you regroup when there are not enough tens or ones.**

Subtract Three-Digit Numbers

Find each difference.

1

357 − 142

	hundreds	tens	ones
	3	5	7
−	1	4	2

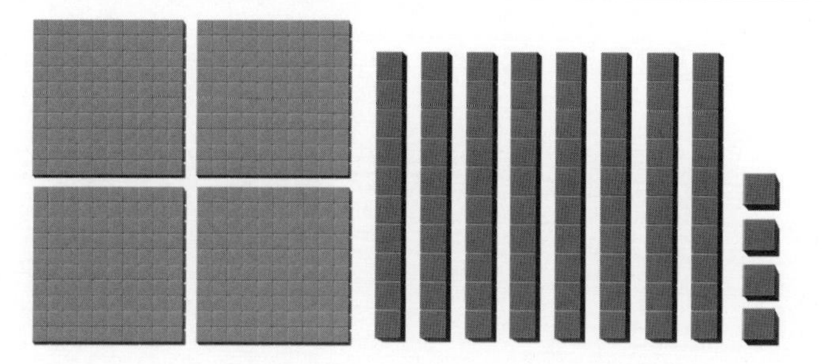

484 − 251

	hundreds	tens	ones
	4	8	4
−	2	5	1

3

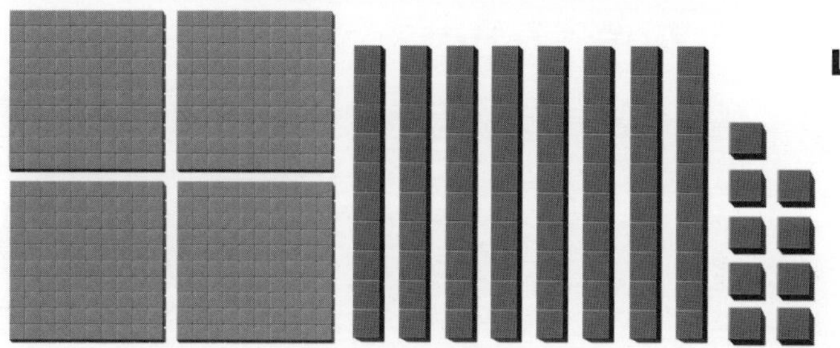

489 − 265

	hundreds	tens	ones
	4	8	9
−	2	6	5

4

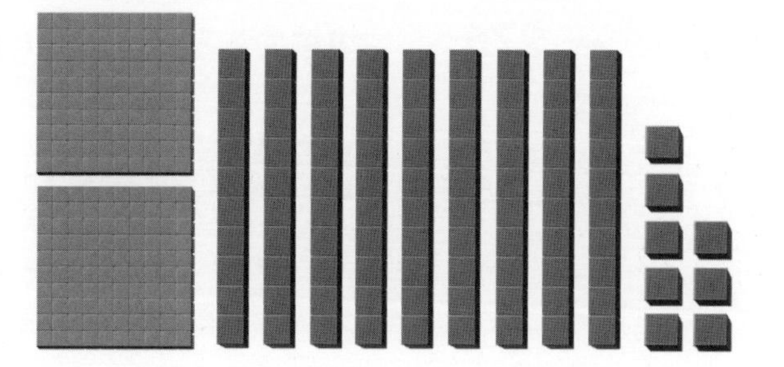

298 − 174

	hundreds	tens	ones
	2	9	8
−	1	7	4

☆ **Tell how you know your answer is reasonable.**

 Name ____________________

Subtract Three-Digit Numbers

Find each difference.

1. 

470 − 247

	hundreds	tens	ones
	4	7	0
−	2	4	7

2. 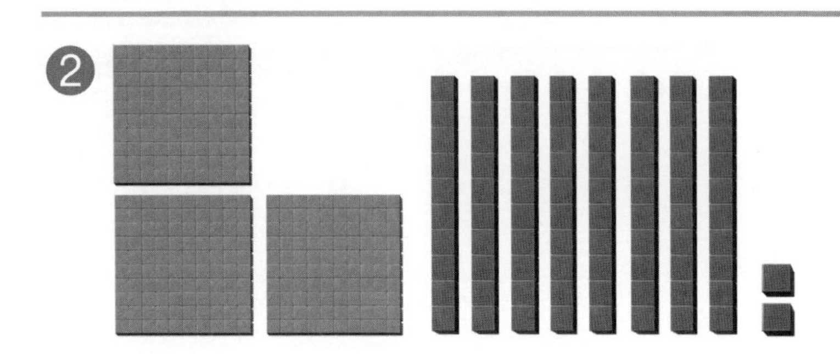

382 − 149

	hundreds	tens	ones
	3	8	2
−	1	4	9

3. 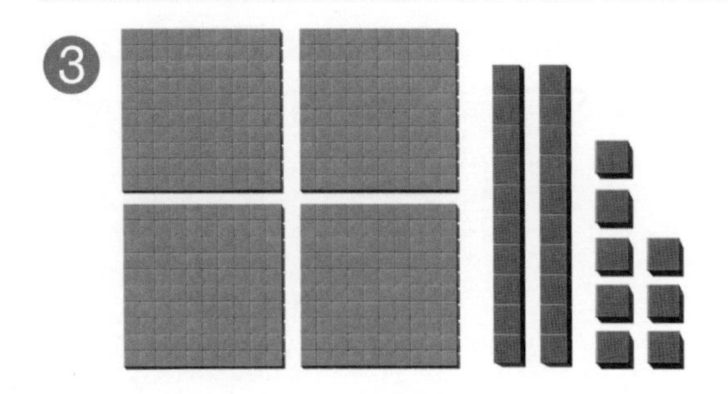

428 − 143

	hundreds	tens	ones
	4	2	8
−	1	4	3

4. 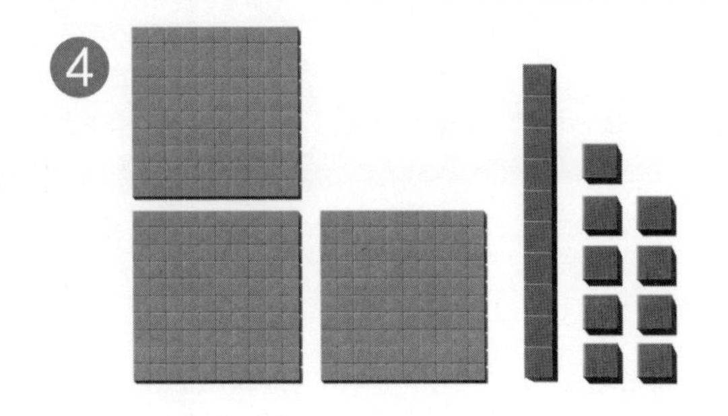

319 − 276

	hundreds	tens	ones
	3	1	9
−	2	7	6

☆ **Tell how you record subtraction with regrouping.**

Subtract Three-Digit Numbers

Find each difference.

1. 967 − 554

	hundreds	tens	ones
	9	6	7
−	5	5	4

2. 791 − 366

	hundreds	tens	ones
	7	9	1
−	3	6	6

3. 648 − 283

	hundreds	tens	ones
	6	4	8
−	2	8	3

4. 821 − 477

	hundreds	tens	ones
	8	2	1
−	4	7	7

☆ **Tell how you regroup to subtract.**

Assessment

Find each difference.

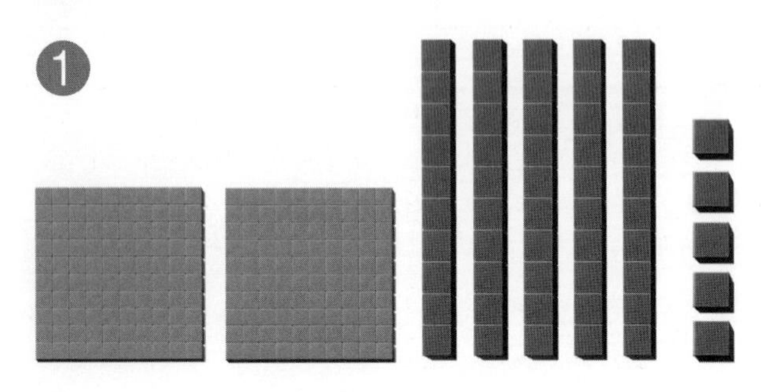

255 – 131

	hundreds	tens	ones
	2	5	5
–	1	3	1

2

362 – 138

	hundreds	tens	ones
	3	6	2
–	1	3	8

786 – 461

	hundreds	tens	ones
	7	8	6
–	4	6	1

4 939 – 275

	hundreds	tens	ones
	9	3	9
–	2	7	5

Tell how you solved the problem.

Overview Inch, Foot, Yard

Directions and Sample Answers for Activity Pages

Day 1	See "Model the Skill" below.
Day 2	Read the directions aloud. Explain that inches and feet are two units of length. Have students measure to the nearest inch or foot. Demonstrate how to mark the end of a ruler and slide the ruler to continue measuring. You may wish to have students use a yardstick or tape measure. Remind students to check that they are recording the measurement on the correct line. Discuss with them how the two measurements relate to the size of the units. Help students to see that a smaller unit has a greater number than a larger unit and that a larger unit has a lesser number than a smaller unit. (Answers will vary.)
Day 3	Read the directions aloud. Have students use their experiences with other measurements to estimate the length of their pencils. Have students use the actual measurements of the objects on the page as they estimate the lengths of the other objects. (Answers will vary.)
Day 4	Read the directions aloud. Point out that students will measure the actual objects in problems 1 and 2 and the pictured objects in problems 3 and 4. Explain that students should measure each object and then compare the measurements to find out how much longer one is than the other. Students might count up or subtract to compare. (Answers: **1.** Answers may vary. Possible answer: about 8 inches; **2.** Possible answer: about 3 feet; **3.** about 2 inches; **4.** about 2 inches)
Day 5	Read the directions aloud. Have students complete the page. Do they align the zero mark of the ruler with one edge of the object? Do they identify the appropriate unit of measure? Do they compare measurements? Use your observations to plan further instruction and review. (Answers: **1.** Answers will vary. **2.** about 6 inches; **3.** about 3 inches; **4.** about 3 inches)

Model the Skill

- Hand out the Day 1 activity page and an inch ruler to each student. Display an inch ruler, a yardstick, and a tape measure, and have students identify each object. **Say:** *These are three different tools that can measure the length of an object. Look at your ruler. What units does a ruler use to measure?* (inches and feet) Discuss that a tape measure can also measure in inches and feet while a yardstick can measure in inches, feet, or yards.
- **Say:** *When you measure an object, line up the zero mark of the ruler with one edge of the object you are measuring. If there is no zero mark, line up the end of the ruler with the edge of the object. Look to see which inch measurement is closest to the other end of the object. About how long is the eraser in problem 1?* (about 2 inches)
- **Say:** *Look at the paintbrush in problem 2. What do you do first to measure the length?* (Place the zero mark of the ruler even with one end of the brush.) *What is the length?* (about 6 inches)
- Help students complete the activity page by measuring the length of the marker. Check that they line up the zero mark of the ruler with the left edge of the object and see which inch the other end of the object is closest to. (Answer: **3.** about 5 inches)

Use Measurement Tools

Use rulers, yardsticks, and tape measures to measure lengths in inches, feet, and yards.

Use each of the tools to measure objects in both inches and feet.

Compare different measures of the same object as well as measures of different objects.

 Name ______________________________

Inch, Foot, Yard

Measure the length of the object.

1

about ______ inches

2

about ______ inches

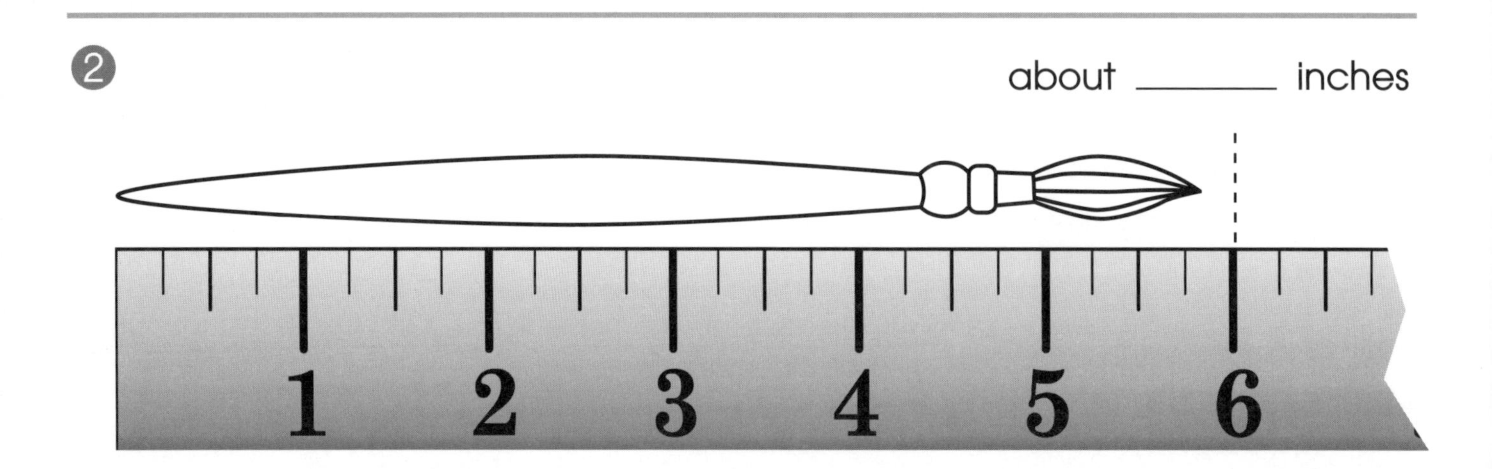

3

about ______ inches

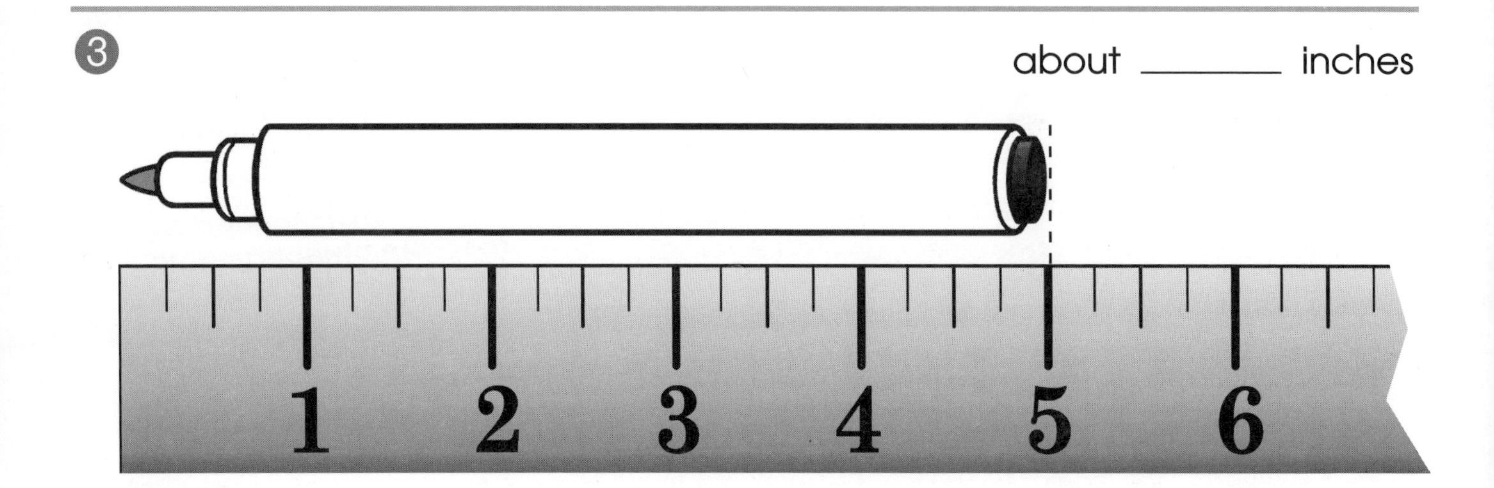

☆ **Tell how you know the length of the object.**

Name ______________________

Inch, Foot, Yard

Measure the length of the object in inches and in feet.

1 your desk

about ______ inches

about ______ feet

2 a door

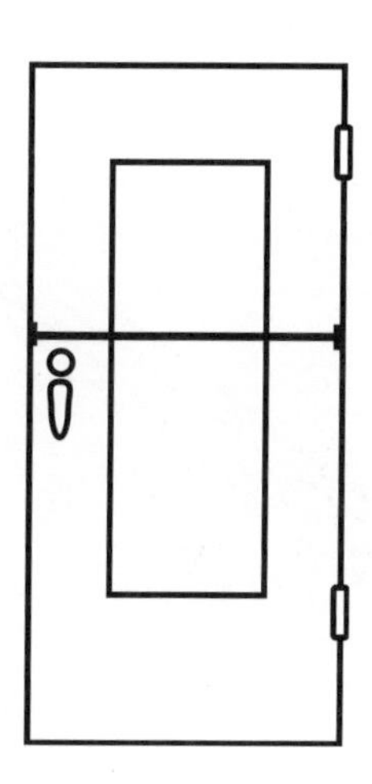

about ______ inches

about ______ feet

3 a board

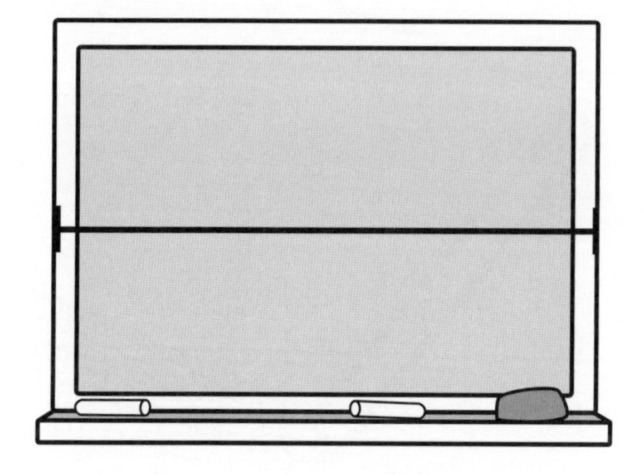

about ______ inches

about ______ feet

4 a classmate

about ______ inches

about ______ feet

☆ **Tell how the two measurements relate to the size of the units.**

 Name ______________________

Inch, Foot, Yard

Estimate the length of the object. Circle the unit. Then measure the object.

1. your pencil

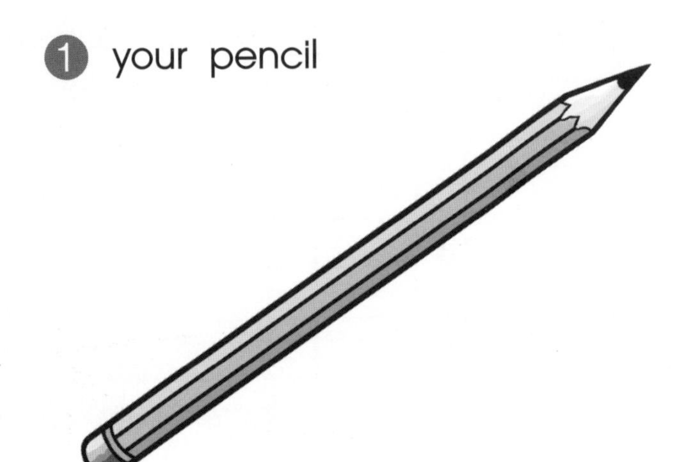

estimate: about ______ inches/feet

measure: about ______ inches/feet

2. a table

estimate: about ______ inches/feet

measure: about ______ inches/feet

3. a crayon

estimate: about ______ inches/feet

measure: about ______ inches/feet

4. a book

estimate: about ______ inches/feet

measure: about ______ inches/feet

☆ **Tell how you know your estimate is reasonable.**

 Name ____________________

Inch, Foot, Yard

Measure the objects to compare their lengths. Circle the unit.

1. How much longer is this activity page than a crayon?

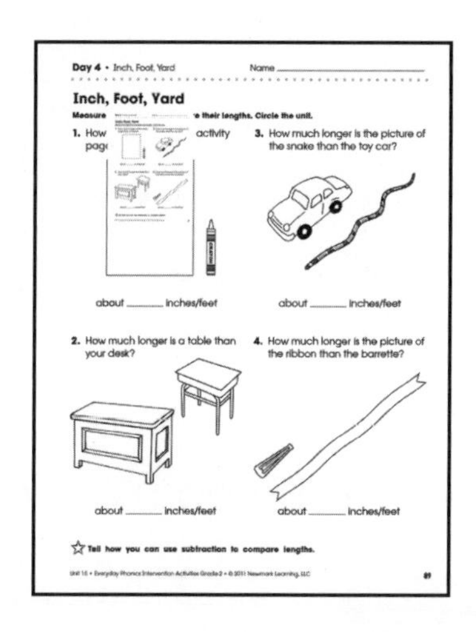
Day 4 • Inch, Foot, Yard Name ____________

Inch, Foot, Yard

3. How much longer is the picture of the snake than the toy car?

about ______ inches/feet about ______ inches/feet

2. How much longer is a table than your desk?

4. How much longer is the picture of the ribbon than the barrette?

about ______ inches/feet about ______ inches/feet

☆ Tell how you can use subtraction to compare lengths.

about ______ inches/feet

2. How much longer is a table than your desk?

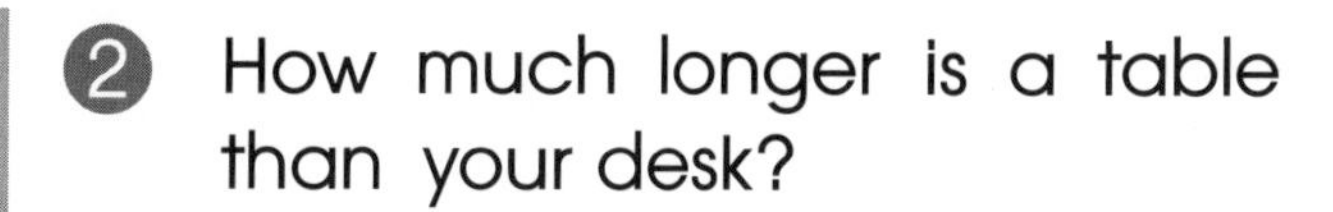

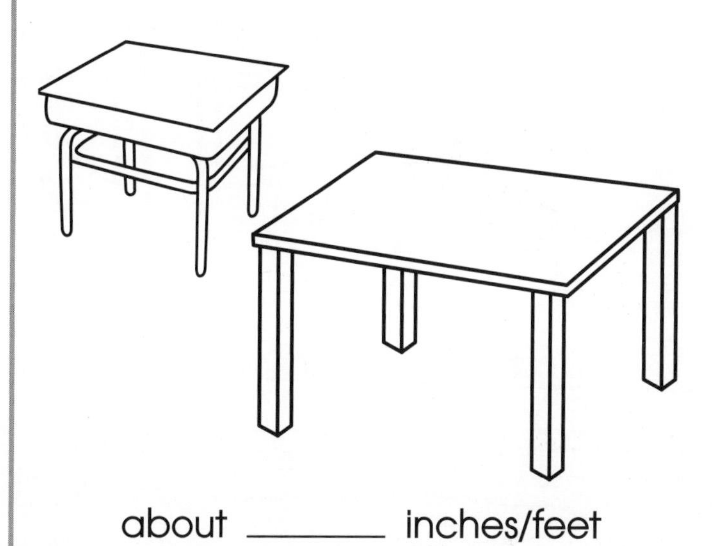

about ______ inches/feet

3. How much longer is the picture of the snake than the toy car?

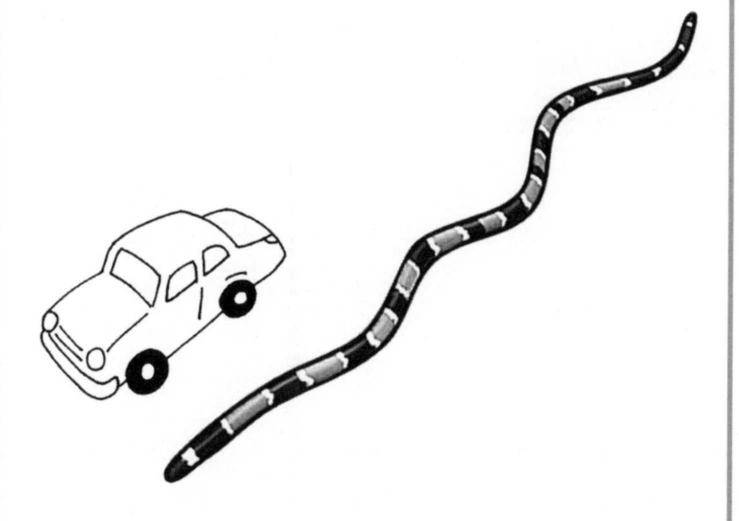

about ______ inches/feet

4. How much longer is the picture of the ribbon than the barrette?

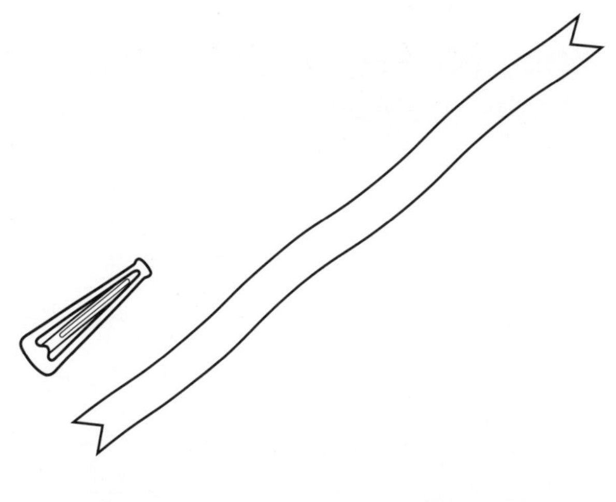

about ______ inches/feet

☆ **Tell how you can use subtraction to compare lengths.**

 Name ______________________________

Inch, Foot, Yard

Solve each problem.

1. Measure the length of your teacher's desk in inches and in feet.

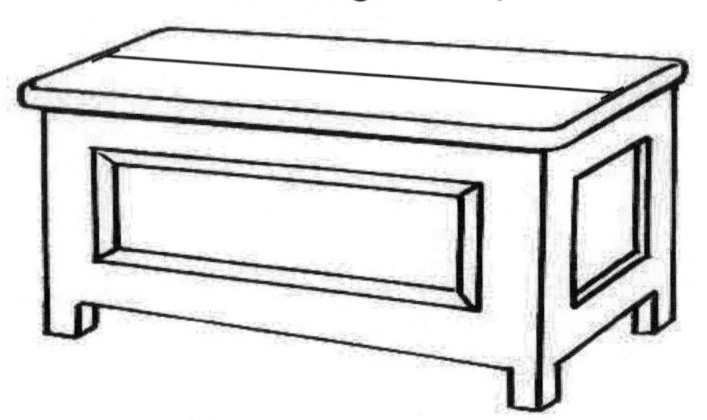

about ________ inches

about ________ feet

2. Estimate the length of this carrot. Circle the unit. Then measure the carrot.

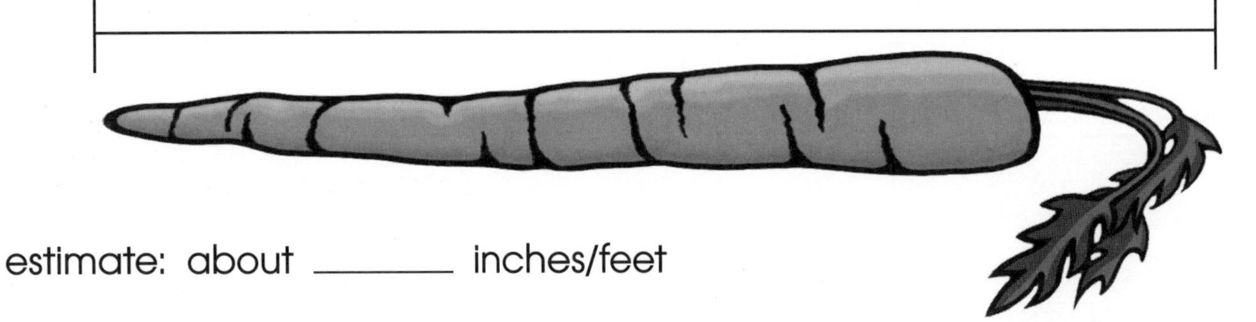

estimate: about ________ inches/feet

measure: about ________ inches/feet

3. Estimate the length of this string bean.
Circle the unit. Then measure the string bean.

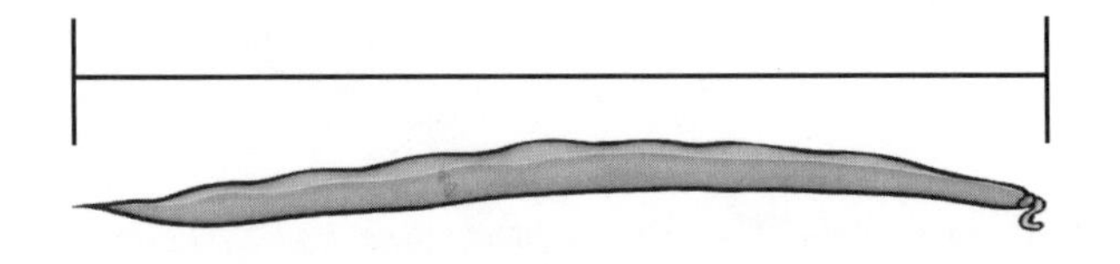

estimate: about ________ inches/feet

measure: about ________ inches/feet

4. How much longer is the picture of the carrot than the picture of the string bean? Circle the unit.

about ________ inches/feet

☆ **Tell how you solved the problem.**

Overview Centimeter, Meter

Directions and Sample Answers for Activity Pages

Day 1	See "Model the Skill" below.
Day 2	Read the directions aloud. Explain that centimeters and meters are two units of length. Direct students to measure to the nearest centimeter or meter. Demonstrate how to mark the end of the ruler and slide it to continue measuring in centimeters, noting that they will need to add measurements to find the total number of centimeters. Discuss with students how the two measurements relate to the size of the units. Guide students to see that a smaller unit has a greater number than a larger unit and that a larger unit has a lesser number than a smaller unit. (Answers will vary. Check that answers are reasonable.)
Day 3	Read the directions aloud. Guide students to use their experiences with other measurements to estimate the length of a chalkboard eraser. Then suggest students use the actual measurements of the objects on the page as they estimate the lengths of the other objects. (Answers will vary. Check that answers are reasonable.)
Day 4	Read the directions aloud. Point out that students will measure pictured objects in problems 1 and 2 and actual objects in problems 3 and 4. Explain that students can measure each object and then compare the measurements, or align the objects and measure the part of the object that does not align. Students might count up or subtract to compare. (Answers: **1.** 6 centimeters; **2.** 8 centimeters; **3.** Answers may vary. Possible answer: 9 centimeters; **4.** Possible answer: 2 meters)
Day 5	Read the directions aloud. Observe as students complete the page. Do students align the zero mark of the ruler with one edge of the object? Do they identify the appropriate unit of measure? Do they accurately compare measurements? Use your observations to plan further instruction and review. (Answers: **1.** Answers will vary. **2.** 13 centimeters; **3.** 7 centimeters; **4.** about 6 centimeters)

Model the Skill

- Hand out the Day 1 activity page and a centimeter ruler to each student. Display a centimeter ruler and a meterstick. **Say:** *These are two different tools that can measure the length of an object in metric units. They look similar to the tools used to measure in customary units—inches, feet, and yards—except they measure in metric units. What are some metric units?* (Possible answers: centimeters, meters, and millimeters) Discuss that a meterstick is the length of 100 centimeters.
- **Say:** *When you measure an object, you need to make sure that you line up the zero mark of the ruler with one edge of the object you are measuring. Then look to see which centimeter measurement is closest to the other end of the object. About how long is the caterpillar in problem 1?* (about 6 centimeters)

Use Measurement Tools

Use metric rulers and metersticks to measure lengths in centimeters and meters.

Compare different measures of the same object as well as measures of different objects.

- **Say:** *Look at the leaf in problem 2. What do you do first to measure the length?* (Place the zero mark of the ruler even with one end of the leaf.) *What is the length?* (about 13 centimeters) Help students complete the activity page by measuring the length of the salamander. Check that students line up the zero mark of the ruler with the left edge of the salamander to see which number the other end is closest to. (about 8 centimeters)

Centimeter, Meter

Measure the length of the object.

1. 

about ______ centimeters

2. 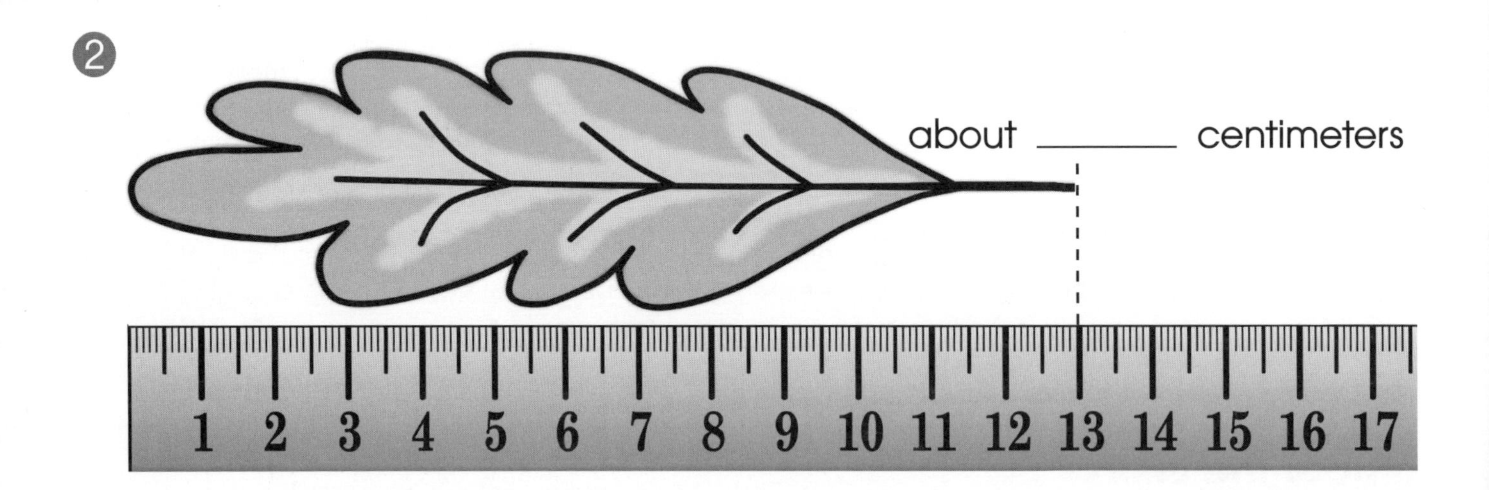

about ______ centimeters

3. 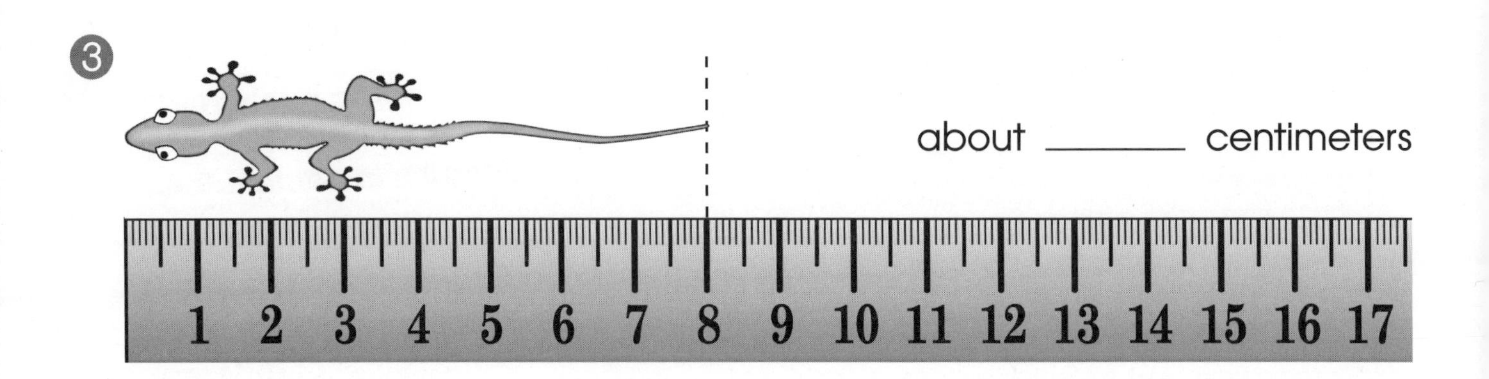

about ______ centimeters

☆ **Tell how you know the length of the object.**

 Name ______________________

Centimeter, Meter

Measure the object in centimeters and in meters.

1 a table

about _______ centimeters

about _______ meters

2 a window

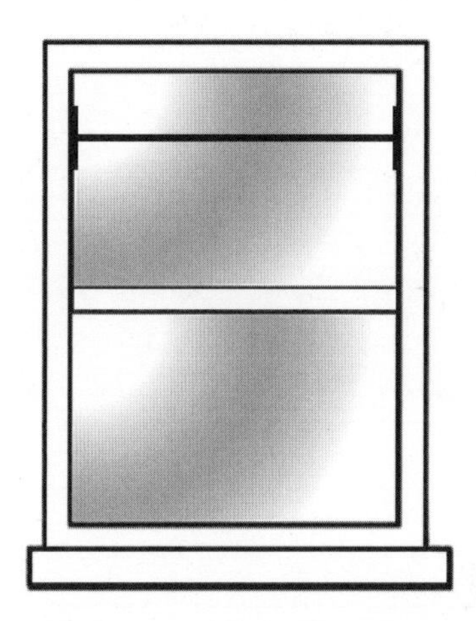

about _______ centimeters

about _______ meters

3 a bulletin board

about _______ centimeters

about _______ meters

4 a bookshelf

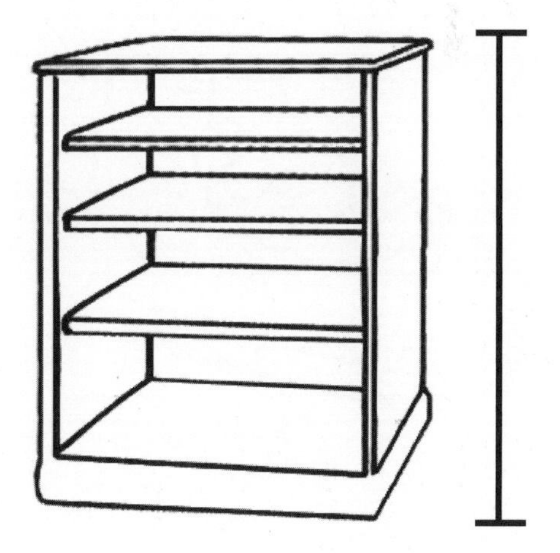

about _______ centimeters

about _______ meters

☆ **Tell how the two measurements relate to the size of the units.**

Name ______________________________

Centimeter, Meter

Estimate the length of the object. Circle the unit. Then measure the object.

1. a board eraser

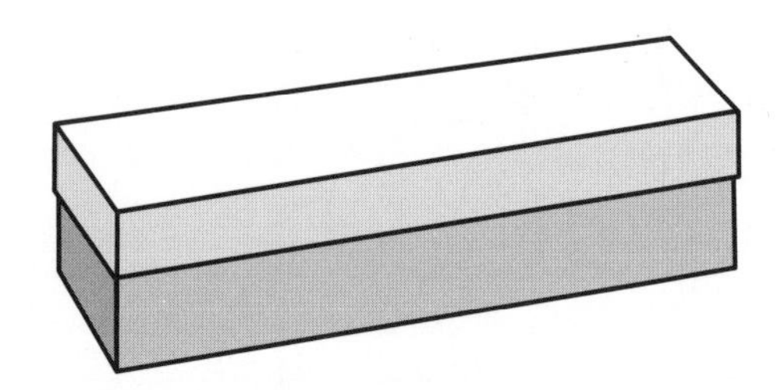

estimate: about ________ centimeters

estimate: about ________ meters

2. a board

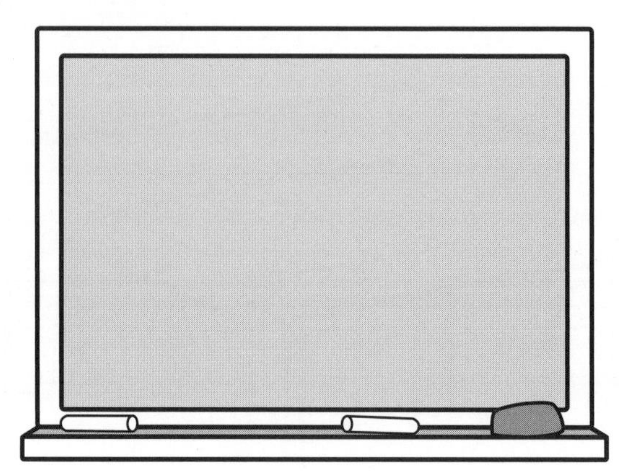

estimate: about ________ centimeters

estimate: about ________ meters

3. your shoe

estimate: about ________ centimeters

estimate: about ________ meters

4. a friend

estimate: about ________ centimeters

estimate: about ________ meters

☆ **Tell how you know your estimate is reasonable.**

 Name ____________________

Centimeter, Meter

Measure the objects to compare their lengths. Circle the unit.

1. How much longer is this name badge than the flag pin?

about ______ centimeters/meters

2. How much longer is this string cheese than the carrot?

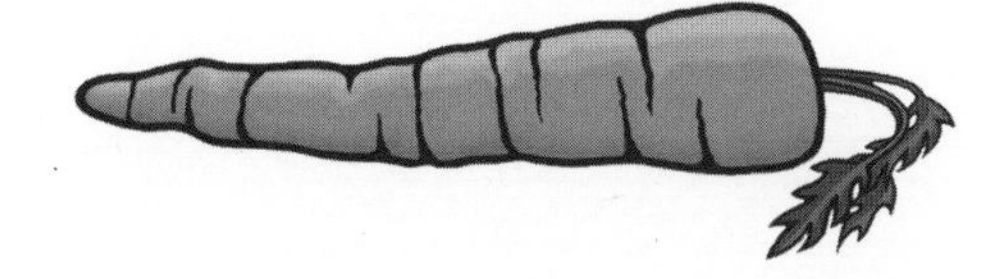

about ______ centimeters/meters

3. How much longer is your pencil than a crayon?

about ______ centimeters/meters

4. How much longer is your teacher's desk than your desk?

about ______ centimeters/meters

 **Tell how you can compare lengths.**

 Name ____________________

Centimeter, Meter

Solve each problem.

1. Measure the classroom closet in centimeters and in meters.

   about ______ centimeters

   about ______ meters

---

2. Estimate the length of this bracelet. Circle the unit. Then measure it.

estimate: about ______ centimeters/meters

measure: about ______ centimeters/meters

---

3. Estimate the length of this toy airplane. Circle the unit. Then measure it.

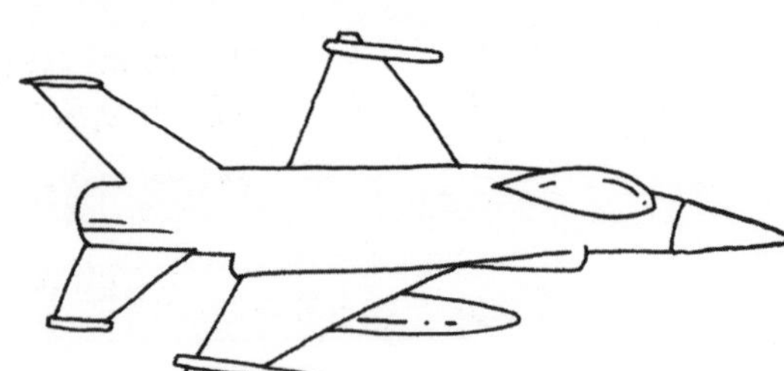

estimate: about ______ centimeters/meters

measure: about ______ centimeters/meters

---

4. How much longer is the bracelet than the toy airplane? Circle the unit.

   about ______ centimeters/meters

☆ **Tell how you solved the problem.**

Overview Add and Subtract Lengths

Directions and Sample Answers for Activity Pages

Day 1	See "Model the Skill" below.
Day 2	Read the directions aloud. Point out that number lines are not given on this page, but that students can draw them to help picture the problem. Explain that students will be using number sentences to solve the word problems on the page. Guide students to see that an open box is used for the unknown number in the number sentence. Students add to find the unknown. (Answers: **1.** 11; **2.** 10 + 9 = 19; **3.** 11 + 4 = 15; **4.** 9 + 8 = 17)
Day 3	Read the directions aloud. Point out that the number line in problem 1 shows subtraction of two lengths.. Have students draw one line for problem 2 and both lines for problem 3. (Answers: **1.** 2; **2.** 6; **3.** 14)
Day 4	Read the directions aloud. Tell students that they will be using number sentences to solve the problems on this page. Remind them that the open box represents the unknown number. Point out that some numbers are shown in the number sentences for problems 1 and 2, but students will need to write the numbers for problems 3 and 4 into the number sentence. Students should subtract to find the unknown in these problems. (Answers: **1.** 9; **2.** 15 – 3 = 12; **3.** 13 – 5 = 8; **4.** 9 – 7 = 2)
Day 5	Read the directions aloud. Have students complete the page. Do students use the number lines to show the operation? Do they write accurate equations to find the unknown? Use your observations to plan further instruction. (Answers: **1.** 14; **2.** 14; **3.** 14 – 5 = 9; **4.** 9 + 7 = 16)

Model the Skill

- Hand out the Day 1 activity page. **Ask:** *What do you see on this page?* (Possible answer: word problems and number lines) Hold up a ruler. **Ask:** *How is a number line like a ruler?* (Possible answer: They both show numbers in order that are equally spaced.) **Say:** *You can use a number line to draw a picture for a word problem.*

Use Number Lines

Use number lines to model two lengths. The second should begin where the first ends.

Add by drawing two lengths to the right.

Subtract by drawing one length to the right and one to the left.

- Read aloud problem 1. **Say:** *Tom used 9 inches of gold wire. Look at the line labeled **gold wire**. It shows a length of 9 units. The problem tells us that he also used 3 inches of silver wire. To find out how many inches of wire he used in all, the 3 inches is added to 9 inches. So there is another line that begins at 9 and shows 3 units. If you look at where the silver wire line ends, you will see the total number of inches that Tom used. What is 9 plus 3?* (12)
- Read aloud problem 2. **Say:** *A line shows the length of purple string as 13 inches. You want to find out how many inches of string Lisa used in all. How should you show the length of the pink string?* (Draw a line 7 units long starting at the 13.) Observe as students draw their lines, offering assistance as needed. **Ask:** *How many inches of string did Lisa use altogether?* (20)
- Help students complete the activity page by drawing lines to show the addition of the two lengths. (Answer: **3.** 18)

 Name ______________________

Add and Subtract Lengths

Add. Use the number line to help you.

1. Tom used 9 inches of gold wire and 3 inches of silver wire. How many inches of wire did he use in all? ______ inches of wire in all

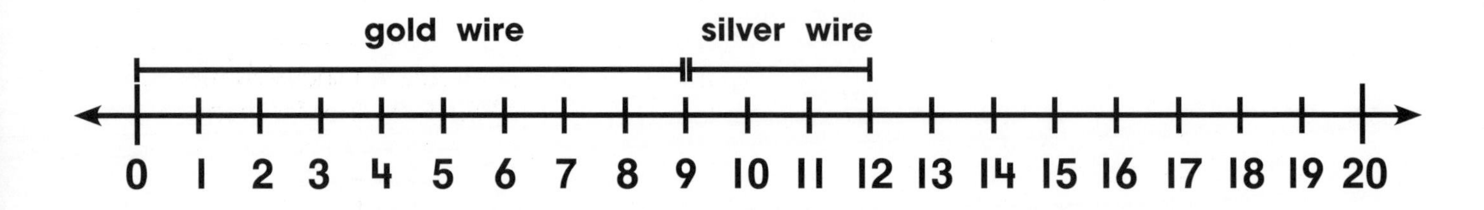

2. Lisa used 13 inches of purple string and 7 inches of pink string. How many inches of string did Lisa use altogether? ______ inches of string

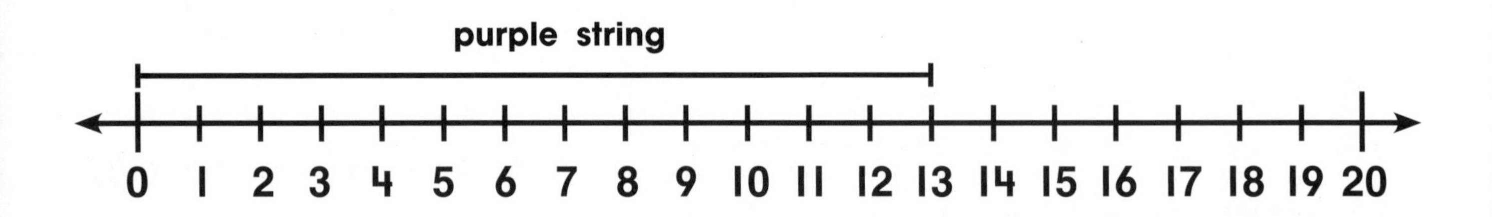

3. Mr. Rocco has a 12-foot piece of garden hose. He bought a new 6-foot piece of garden hose. How many feet of garden hose does Mr. Rocco have now?

______ feet of garden hose

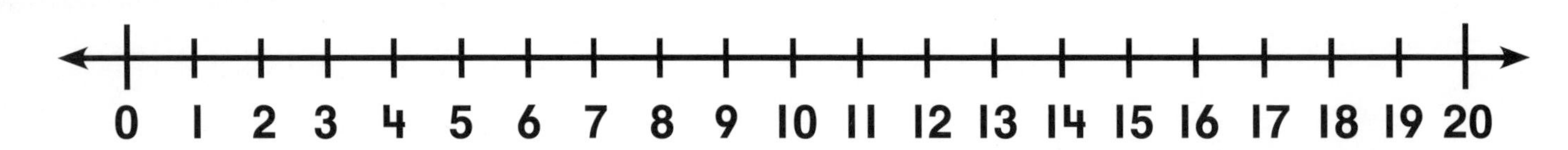

☆ **Tell how you can use a number line to add lengths.**

 Name ______________________

Add and Subtract Lengths

Add. Complete the number sentence.

1. Donna has 6 yards of black cord and 5 yards of red cord.

   How many yards of cord does Donna have in all?

   6 + 5 = ☐

   ______ yards of cord

2. Jim put up a 10-foot ladder.

   He extended it 9 feet.

   How many feet does the ladder now reach?

   10 + ______ = ☐

   ______ feet

3. Scott had an 11-foot piece of carpet.

   He bought 4 more feet of carpet.

   How many feet of carpet does he have now?

   ______ + ______ = ☐

   ______ feet of carpet

4. Barbara put a 9-inch piece of trim on a hat.

   Then she put on a piece of 8-inch trim.

   How many inches of trim did Barbara put on the hat?

   ______ + ______ = ☐

   ______ inches of trim

☆ **Tell how you know the unknown number in the problem.**

 Name ______________________________

Add and Subtract Lengths

Subtract. Use the number line to help you.

1. Mark had 10 meters of tape. He used 8 meters of tape. How many meters of tape does Mark have left?

______ meters of tape left

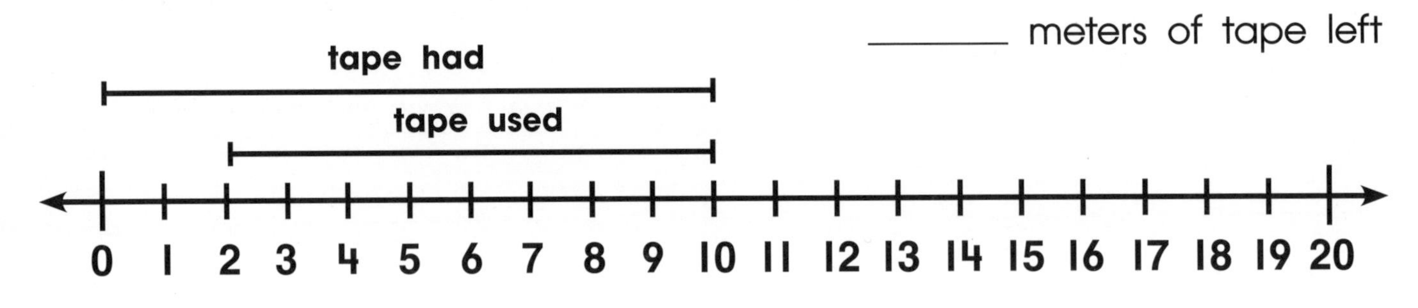

2. Keith bought 18 feet of trim. He used 12 feet. How many feet of trim does Keith have left?

______ feet of trim left

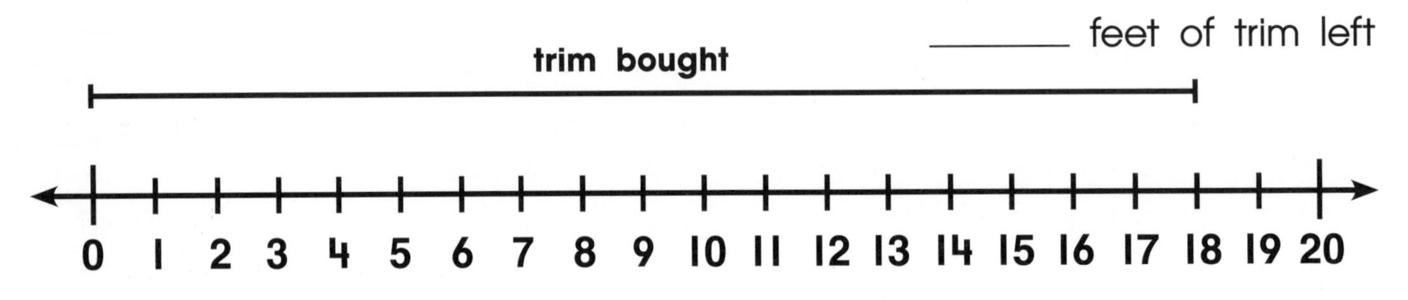

3. A ball of string has 18 yards on it. If Rachel uses 4 yards, how many yards of string will be left?

______ yards of string

0 1 2 3 4 5 6 7 8 9 10 11 12 13 14 15 16 17 18 19 20

Add and Subtract Lengths

Subtract. Complete the number sentence.

1. The driveway is 19 feet long.

   The family paved 10 feet.

   How many feet were left unpaved?

   19 – 10 = ☐

   ______ feet left unpaved

2. Tom used a 15-centimeter strip of paper as a bookmark.

   He cut off 3 centimeters.

   How long is the bookmark now?

   15 – ______ = ☐

   ______ centimeters

3. Mia's hair was 13 inches long.

   She had 5 inches cut off.

   How long is Mia's hair now?

   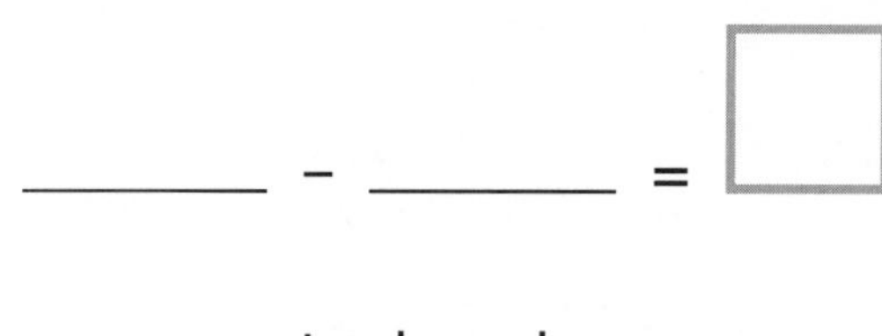

   ______ – ______ = ☐

   ______ inches long

4. Miss Black bought 9 yards of ribbon.

   She used 7 yards to make a wreath.

   How many yards of ribbon does Miss Black have left?

   ______ – ______ = ☐

   ______ yards of ribbon

☆ **Tell how you solve the problem to find the unknown.**

 Name ____________________

Assessment

Solve each problem. Show your work.

1. Mr. Hanks needs 8 feet of paper to cover one bulletin board and 6 feet of paper to cover another bulletin board. How many feet of paper does Mr. Hanks need to cover the two bulletin boards?

_______ feet of paper

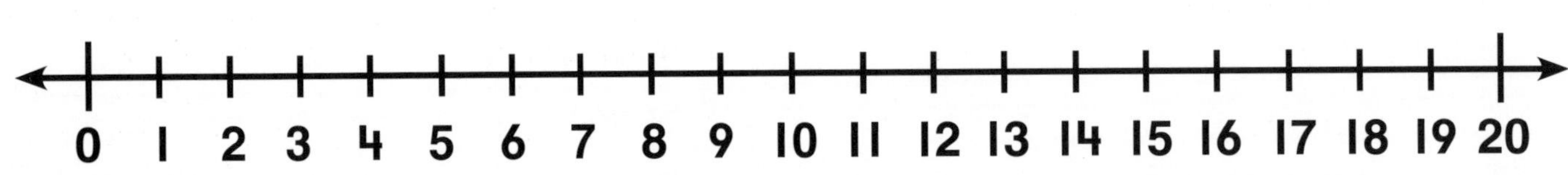

2. Gina bought 20 yards of fabric. She used 6 yards to make curtains. How many yards of fabric does Gina have left?

_______ yards of fabric

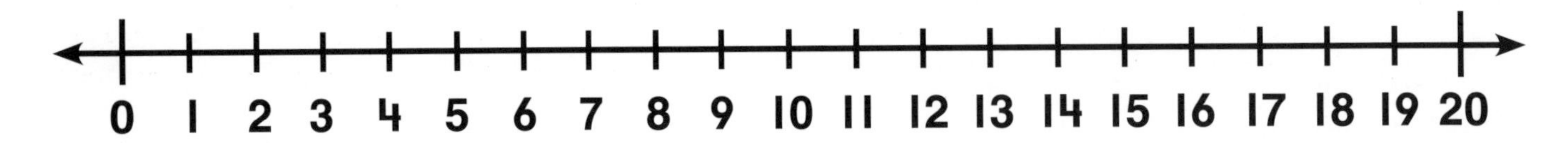

3. Amy had a 14-inch piece of sequin trim.

She used 5 inches on her costume.

How much sequin trim does Amy have left?

_______ − _______ = ☐

_______ inches of sequin trim

4. Charlie has 9 centimeters of black wire and 7 centimeters of red wire.

How much wire does Charlie have in all?

_______ + _______ = ☐

_______ centimeters of wire

☆ **Tell how you solved the problem.**

Overview Tell Time to the Nearest Five Minutes

Directions and Sample Answers for Activity Pages

Day 1	See "Model the Skill" below.
Day 2	Read the directions aloud. Point out that each analog clock in the first column has a matching digital clock in the second column. (Answers: **1.** 1:30; **2.** 4:30; **3.** 11:30; **4.** 8:30)
Day 3	Read the directions aloud. Have students differentiate between the hour hand and the minute hand. Remind students that five minutes pass between each number on the clock. Have students count by fives as they point to each number. Point out that the colon is given so students need to write the hour before the colon and the minutes after the colon. Remind students that the hour is the number that the hour hand has passed. Encourage students to count by fives to determine the minutes. (Answers: **1.** 12:25; **2.** 6:10; **3.** 10:35; **4.** 3:50)
Day 4	Read the directions aloud. Guide students to use the sky or the activity in the pictures to help them determine whether the time shown is A.M. or P.M. Discuss with students that A.M. times are from midnight to noon and P.M. times are from noon to midnight. (Answers: **1.** 8:30 P.M.; **2.** 8:25 A.M.; **3.** 7:00 A.M.; **4.** 3:45 P.M.)
Day 5	Read the directions aloud. Have students complete the page. Do students identify the correct hour? Do they identify the correct minutes? Do they identify a time as A.M. or P.M.? Use your observations to plan further instruction. (Answers: **1.** 2:00; **2.** 7:30; **3.** 2:55; **4.** 8:20 A.M.)

Model the Skill

- Hand out the Day 1 activity page. Draw an analog clock that shows 10 o'clock. **Say:** *You can show the same time in more than one way. What time is shown on this clock?* (10 o'clock) Write **10 o'clock** on the board. **Say:** *There is another way to write* ***10 o'clock*** *with only numbers.* Write **10:00**. **Say:** *The numbers before the colon tell the hour and the numbers after the colon tell the minutes.* Point out that 10:00 says hour 10 and zero minutes.

- **Say:** *Look at problem 1. Times are shown in three different ways. Which two times are the same?* (the two clocks) Have students circle the two clocks. **Ask:** *Why are the words 12 o'clock not like the two clocks?* (The clocks show 2:00 and the words show 12:00.) Have students cross out the words **12 o'clock**.

- **Ask:** *Look at problem 2. What time is shown on the digital clock?* (5 o'clock) *Is that the time shown with the words?* (yes) *Does the circular clock show 5 o'clock?* (no) Guide students to circle the digital clock and the words and to cross out the analog clock.

- Help students complete the activity page by deciding if either clock matches the words or if the two clocks both show the same time. Have students circle two times and cross out one. (Answers: **3.** Students should circle the analog clock and the words, and cross out the digital clock. **4.** Students should circle the two clocks, and cross out the words.

 Name ______________________

Tell Time to the Nearest Five Minutes

Circle the two times that are the same. Cross out the time that is different.

1 **12 o'clock**

2:00

2 **5 o'clock**

5:00

3 **9 o'clock**

12:00

4 **7 o'clock**

6:00

☆ **Tell how you know what time is shown.**

Tell Time to the Nearest Five Minutes

Draw lines to match clocks that show the same times.

☆ **Tell how you know what time is shown.**

Tell Time to the Nearest Five Minutes

Write each time.

1

_______ : _______

2

_______ : _______

3

_______ : _______

4

_______ : _______

☆ **Tell how you tell time to the nearest five minutes.**

Tell Time to the Nearest Five Minutes

Write the time shown. Circle A.M. or P.M.

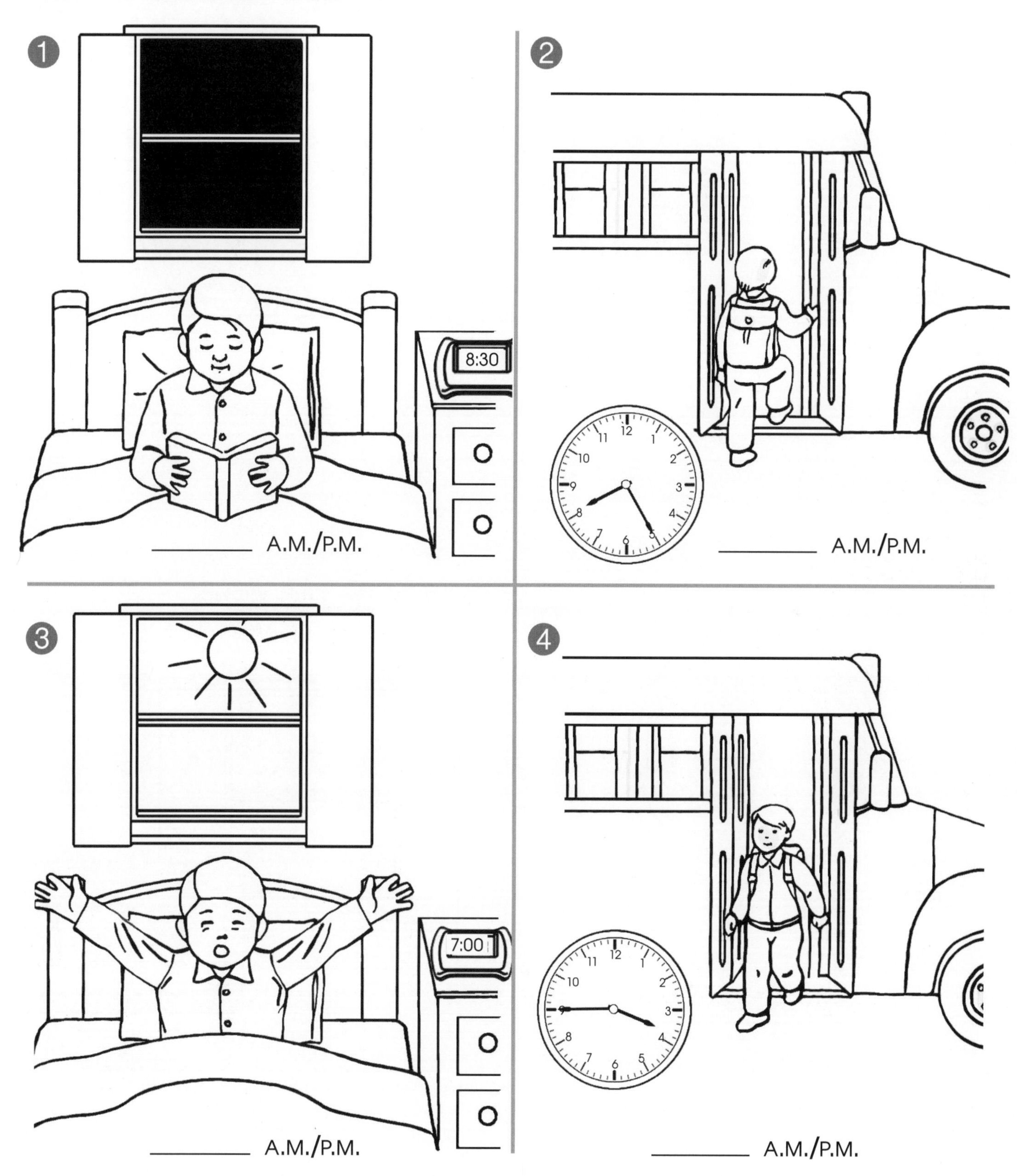

☆ **Tell how you know a time is A.M. or P.M.**

 Name ______________________

Assessment

Write each time.

1

____

2

____

3

____

4 **Write the time shown.**
Circle A.M. or P.M.

______________________ A.M./P.M.

☆ **Tell how you solved the problem.**

Overview How Much Money?

Directions and Sample Answers for Activity Pages

Day 1	See "Model the Skill" below.
Day 2	Read the directions aloud. Point out that students will need to write the cents symbol for each answer on this page. Review with students the values of each coin and how to use skip counting to count dimes and nickels. Suggest that students draw a picture to solve problem 4, counting the dimes first, and then counting on the nickels and pennies. (Answers: **1.** 21¢; **2.** 28¢; **3.** 26¢; **4.** 32¢)
Day 3	Read the directions aloud. Point out that two price tags use a cents symbol and two price tags use a dollar sign. Review with students how to read the prices written with a dollar sign. Encourage students to count on from the coin with the greatest value to the coin with the least value—quarter, dime, nickel, and penny. (Answers: **1.** 27¢; **2.** \$0.45; **3.** \$0.30; **4.** 42¢)
Day 4	Read the directions aloud. Review with students how the dollar amount is written, followed by a decimal point, and then the cents amount. Guide students to use a zero as a placeholder in the answer for problems 3 and 4. (Answers: **1.** \$1.50; **2.** \$1.45; **3.** \$2.05; **4.** \$2.03)
Day 5	Read the directions aloud. Observe as students complete the page. Do students include all the coins in their total? Do they use skip counting? Do they count the coins with the greatest value first? Do they write the amounts with the correct symbols? Use your observations to plan further instruction. (Answers: **1.** 17¢ or \$0.17; **2.** 21¢ or \$0.21; **3.** 80¢ or \$0.80; **4.** \$2.07)

Model the Skill

- Hand out the Day 1 activity page and coins. **Ask:** *What kinds of coins are shown in problem 1?* (nickels and pennies) *What is the value of a nickel?* (5 cents) **Say:** *Place nickels on the pictured nickels and count the values with me—5, 10.* Observe as students place the nickels. **Say:** *Now take pennies and count on from the 10—11, 12, 13.* Observe as students place the pennies. **Ask:** *What is the total amount shown in problem 1?* (13 cents) Point out that the cents symbol is shown so students only need to write the numeral 13.
- **Ask:** *What kinds of coins are shown in problem 2?* (dimes and pennies) *What is the value of a dime?* (10 cents) *How do you count a set of dimes?* (by tens) Have students place dimes on the pictured dimes and count by tens. Have them place the pennies. **Ask:** *What is the total amount shown in problem 2?* (33 cents)
- **Say:** *For problem 3, you need to write the cents symbol. Will you skip count by fives or tens to find the total amount?* (skip count by tens) Observe as students complete problem 3. (41¢)
- Help students complete problem 4 by showing the coins listed, counting the nickels by fives and counting on the penny. (26¢)

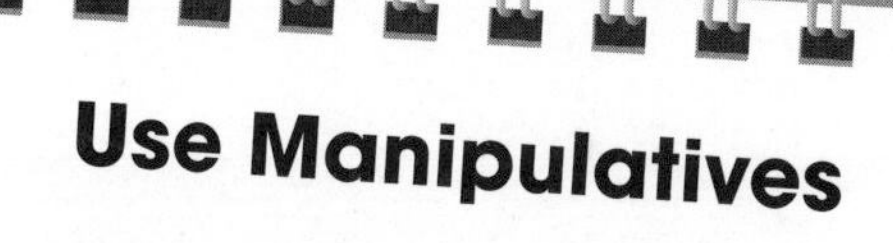

Use Manipulatives

Show a set of mixed coins.

Have students arrange the coins in order from greatest value to least value.

Use skip counting to find the total value of the set of coins.

Name ______________________

How Much Money?

Write each amount. Use the ¢ symbol.

1

__________¢

2

____

3

__________¢

4 You have 5 nickels and I penny.

How many cents do you have?

____

☆ **Tell how you can use skip counting to find amounts of money.**

 Name ____________

How Much Money?

Write each amount. Use the ¢ symbol.

1

____

2

____

3

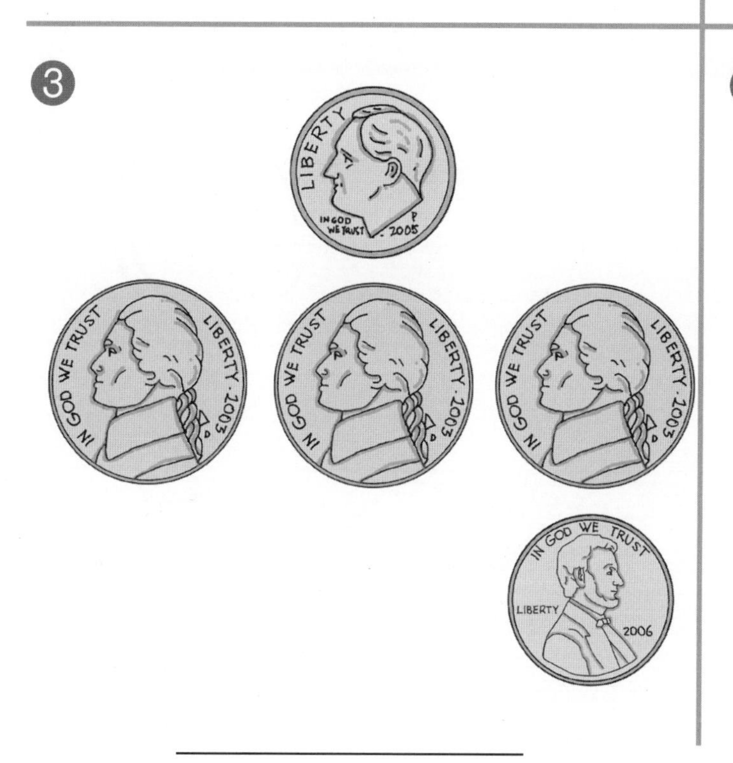

____

4 You have 2 dimes, 2 nickels, and 2 pennies.

What amount of money do you have?

____

☆ **Tell how you count three different kinds of coins.**

 Name ____________________

How Much Money?

Match each purse to a price tag.

1

2

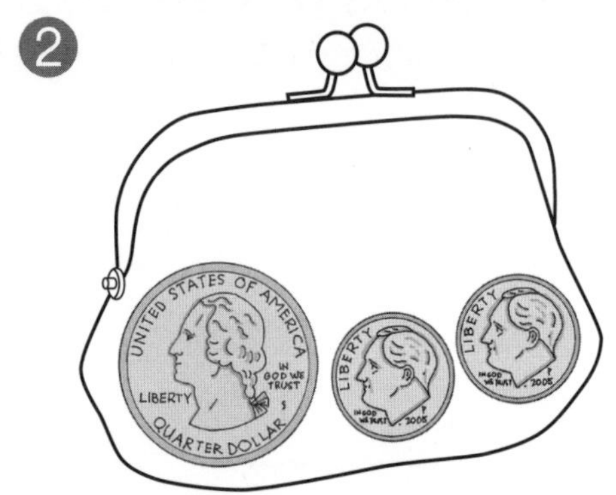

3

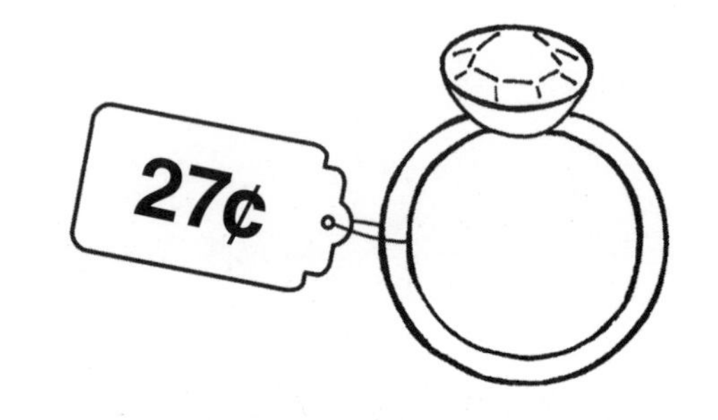

4

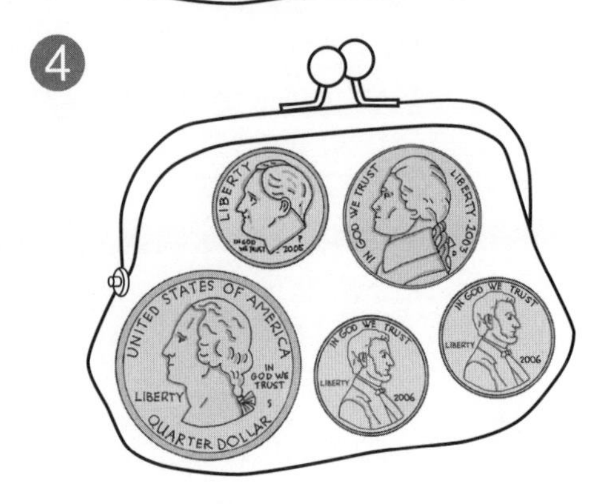

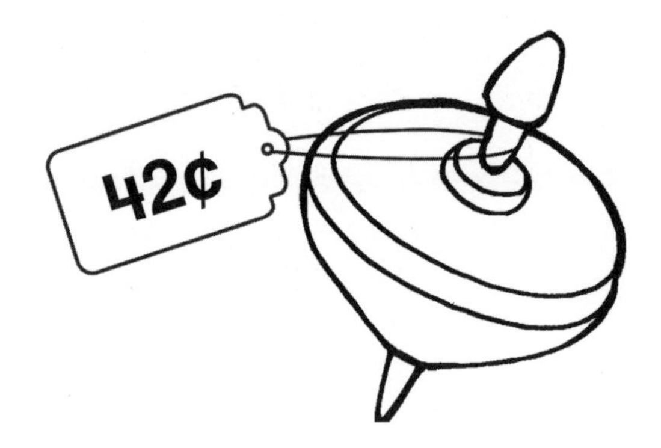

☆ **Tell how you count on to find the total amount of money.**

How Much Money?

Write each amount. Use the $ symbol.

____

____

3

____

4 Sam has two one-dollar bills and **3** pennies. How much money does Sam have?

____

☆ **Tell how you write an amount with dollars and cents.**

 Name ____________________

Assessment

Write each amount. Remember to use symbols.

1. 

____

2. 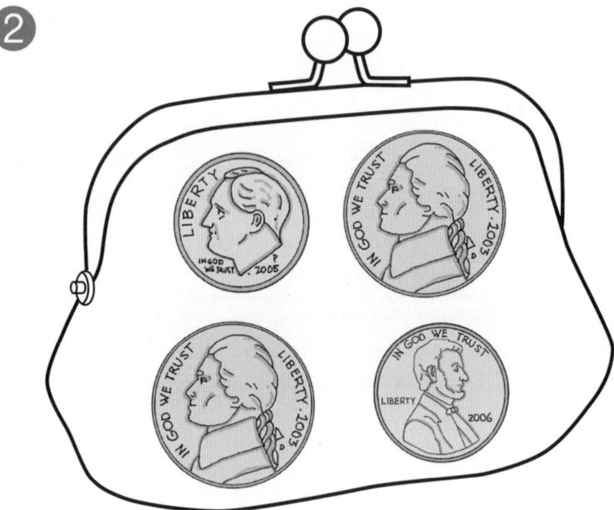

____

3. Molly has three quarters and one nickel.

How much money does Molly have?

____

4. You have two one-dollar bills, a nickel, and two pennies.

What amount of money do you have?

____

☆ **Tell how you solved the problem.**

Overview Make a Line Plot

Directions and Sample Answers for Activity Pages

Day 1	See "Model the Skill" below.
Day 2	Read the directions aloud. Point out the data shown in the tally charts and the parts of the line plot that are given. Guide students to see that the number of Xs above each number should match the number of tally marks. After students complete the line plots, ask questions about the data. (Answers: **1.** Students should show 4 Xs above the 5, 8 Xs above the 6, and 3 Xs above the 7. **2.** Students should show 3 Xs above the 4, 9 Xs above the 5, 6 Xs above the 6, and 2 Xs above the 7.)
Day 3	Read the directions aloud. Point out that the data is given in a list and that students need to graph the data on the line plots. Guide students to cross out each piece of data as they graph it. (Answers: **1.** Students should show 3 Xs above the 4, 6 Xs above the 5, and 4 Xs above the 6. **2.** Students should show 2 Xs above the 5, 5 Xs above the 6, 3 Xs above the 7, and 7 Xs above the 8.)
Day 4	Read the directions aloud. Point out that the data is given in a tally chart in problem 1 and as a list in problem 2. Explain that students need to label the scale and the horizontal axis as well as plot the data. Ask questions about the data on the line plots. (Answers: **1.** Students should label the scale 6–9 and the axis "Length in Inches," and show 2 Xs above the 6, 3 Xs above the 7, 5 Xs above the 8, and 4 Xs above the 9. **2.** Students should label the scale 7–10, the axis "Length in Inches," and show 1 X above the 7, 2 Xs above the 8, 4 Xs above the 9, and 5 Xs above the 10.)
Day 5	Read the directions aloud. Observe as students complete the page. Do students label the scale and the axis? Do they plot the data accurately? Use your observations to plan further instruction and review. (Answers: **1.** Students should show 2 Xs above the 15, 3 Xs above the 16, 6 Xs above the 17, and 4 Xs above the 18. **2.** Students should label the scale 15–18 and the axis "Lengths in Inches," and show 3 Xs above the 15, 2 Xs above the 16, 1 X above the 17, and 4 Xs above the 18.)

Model the Skill

- Hand out the Day 1 activity page. **Say:** *Data can be shown in charts and graphs. A line plot shows data on a number line. The data is plotted with Xs above the values on the scale to show how many of each value.* Look at the line plot on the page. **Ask:** *What does this line plot show?* (the heights of the students in Mrs. Smith's class) *What are the different heights of the students?* (44, 45, 46, 47, and 48 inches) Have students point to the scale that shows the heights and see the labels.

- **Say:** *The Xs show how many students are each height. How can you find the answer to problem 1—the most common height of the students in the class?* (Look for the height that has the most Xs.) **Ask:** *Which height has the most Xs?* (46 inches) **Say:** *Most of the students are 46 inches tall.*

- Look at problem 2. **Ask:** *How do you find the number of students that are 45 inches tall?* (Look for the 45 on the scale and count the number of Xs above the 45.) *What number did you count?* (3)

- Help complete the activity page by using the scale to find the measurements and counting the Xs to find the number of students. (Answers: **3.** 2; **4.** 47 inches)

Use Measurement

Have students use centimeter rulers to measure lengths of objects, such as pencils.

Record the class data in a chart.

Have volunteers record the data on a line plot.

Name ______________________________

Make a Line Plot

Solve each problem. Use the data in the line plot.

The line plot shows the heights
of the students in Mrs. Smith's class.

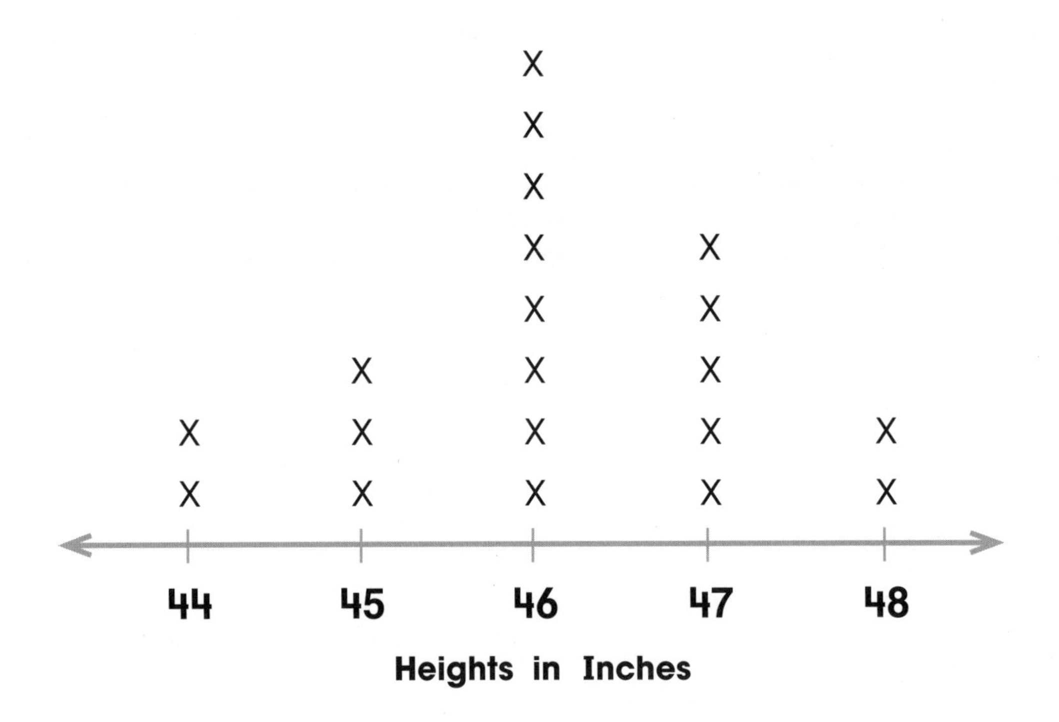

1. What is the most common height of the students in the class? ______

____

2. How many students are 45 inches tall? ______

____

3. How many students are 48 inches tall? ______

____

4. Which height shows 5 students? ______

☆ **Tell how you read a line plot.**

 Name ____________________

Make a Line Plot

Use the data in the tally chart to complete the line plot.

Seedlings in Mr. Gomez's class	
Height in Centimeters	**Number**
4	𝍸
5	IIII
6	𝍸 III
7	III

This line plot shows the height of the seedlings in Mr. Gomez's class.

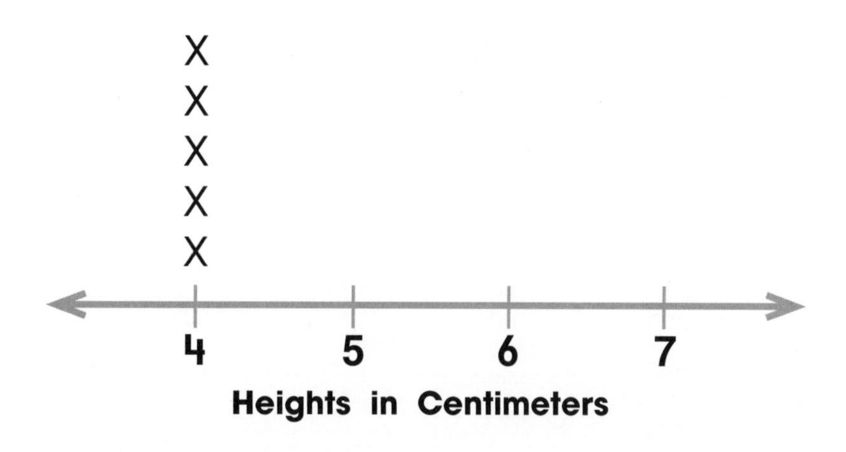

Seedlings in Miss Harvey's class	
Height in Centimeters	**Number**
4	III
5	𝍸 IIII
6	𝍸 I
7	II

This line plot shows the height of the seedlings in Miss Harvey's class.

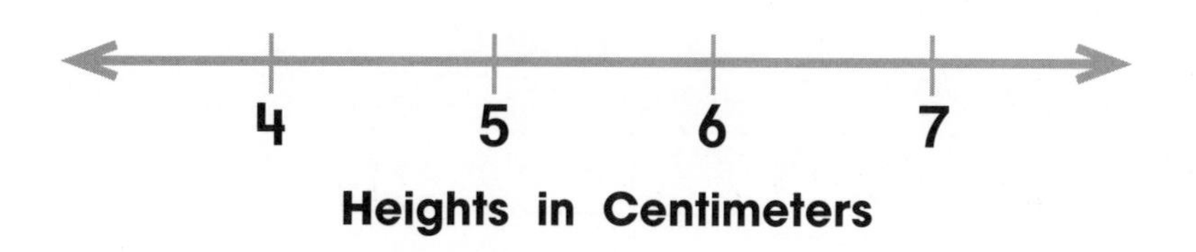

☆ **Tell how you know how many Xs to plot for each value.**

 Name ______________________________

Make a Line Plot

Solve each problem.

1. Mrs. King measured the height of the sunflowers in her garden. She made a list of the height. Use the data to complete the line plot.

Height of Sunflowers

4 feet	4 feet	5 feet
5 feet	3 feet	6 feet
3 feet	4 feet	6 feet
3 feet	5 feet	5 feet
3 feet	6 feet	5 feet
5 feet	3 feet	6 feet

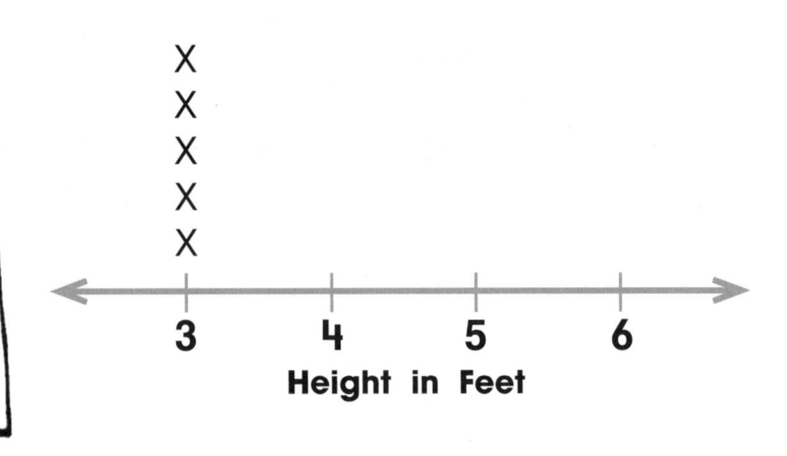

2. Mr. King measured the height of the corn stalks in his garden. He made a list of the height. Use the data to complete the line plot.

Height of Corn Stalks

8 feet	7 feet,	5 feet
8 feet	8 feet	6 feet
8 feet	7 feet	6 feet
8 feet	7 feet	8 feet
6 feet	5 feet	6 feet
8 feet	6 feet	

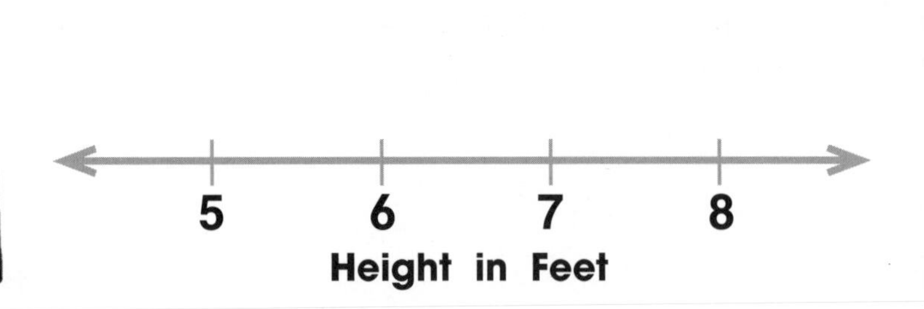

☆ **Tell how you know that you have plotted all the data.**

Name ______________________________

Make a Line Plot

Use the data to complete the line plot.

1. The list shows the length of the carrots Tina peeled. Make a line plot of the data.

Carrots Tina Peeled	
Length in Inches	Number
6	II
7	III
8	~~IIII~~
9	IIII

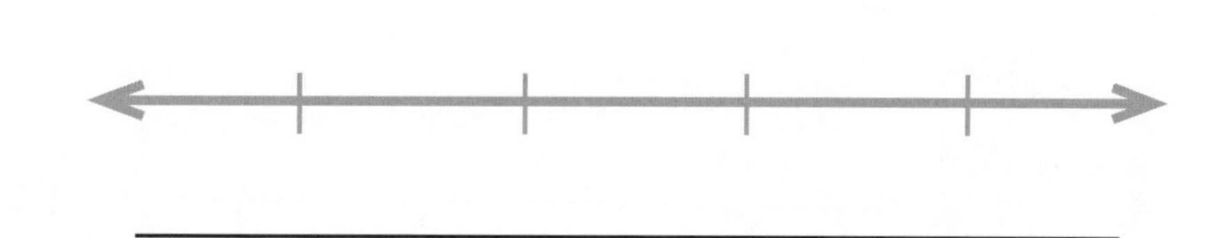

____

2. The list shows the length of the corn cobs that Roger chose for dinner. Make a line plot of the data.

Length of Corn Cobs	
8 inches	10 inches
9 inches	10 inches
10 inches	9 inches
8 inches	7 inches
9 inches	10 inches
9 inches	10 inches

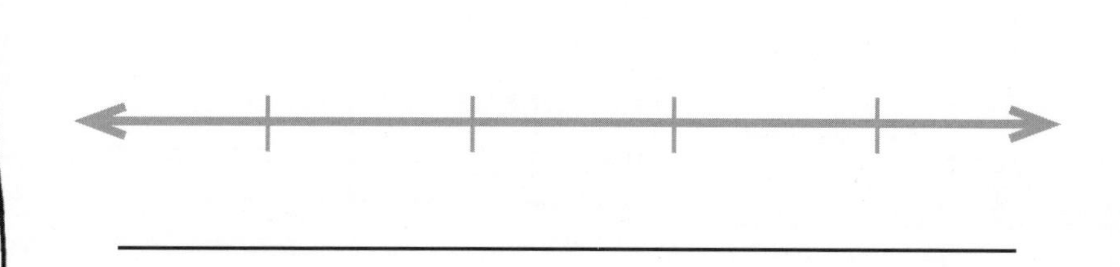

____

☆ **Tell how you know what to label the scale.**

Name ______________________________

Assessment

Use the data to complete the line plot.

Bracelets Patty Made	
Length in Centimeters	**Number**
15	||
16	|||
17	~~||||~~ |
18	||||

This line plot shows the lengths of the bracelets Patty made.

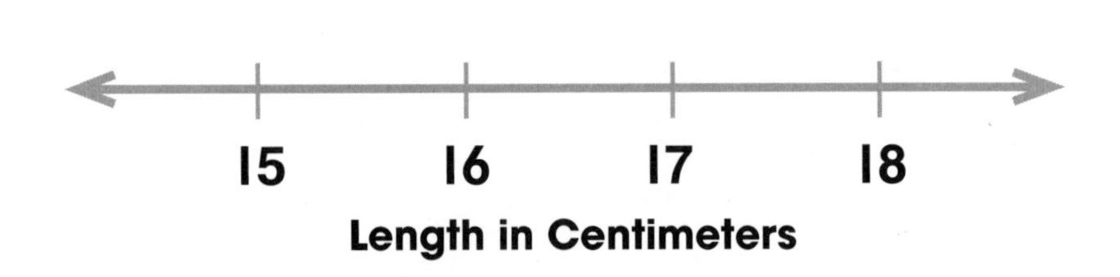

❷ The list shows the length of the necklaces that Morgan made.

Length of Necklaces	
15 inches	17 inches
16 inches	18 inches
18 inches	16 inches
15 inches	18 inches
18 inches	15 inches

____

☆ **Tell how you solved the problem.**

Overview Make a Graph

Directions and Sample Answers for Activity Pages

Day 1	See "Model the Skill" below.
Day 2	Read the directions aloud. Review the parts of the graph and the data shown. Point out that the operation symbol is given only in problem 1 and students will have to decide to add or subtract to solve the other problems. (Answers: **1.** 7; 19 – 7 = 12; **2.** 4; 5; 4 + 5 = 9; **3.** 7 – 4 = 3)
Day 3	Read the directions aloud. Help students see that they only have to draw the bars for the graph in problem 1 and that they need to complete more parts of the graph in problem 2. When the graphs are complete, ask questions about the data on the graphs. (Answers: **1.** Students should draw bars as follows: Soccer 6, Baseball 2, Basketball 4, and Football 7. **2.** Students should title the graph "Pets We Have"; label the horizontal axis "Pets" and list Dog, Cat, and Fish; label the vertical axis "Number of Students" and the scale 0–10; and draw bars—Dog 10, Cat 8, and Fish 2.)
Day 4	Read the directions aloud. Explain that the graphs on this page use pictures to show values. Point out the key. Help students see how the data from the chart is reflected in the graph. Explain that they can use the graph in problem 1 to help them complete the graph in problem 2. (Answers: **1.** Students should draw 7 smiling faces for Blue, 3 for Yellow, and 6 for Green. **2.** Students should title the graph "Favorite Snacks" with the category labels Pretzels, Yogurt, and Fruit. They should designate a symbol in the key and draw 5 for pretzels, 3 for yogurt, and 8 for fruit.)
Day 5	Read the directions aloud. Observe as students complete the page. Do students complete the graphs and correctly solve the word problems? Use your observations to plan further instruction. (Answers: **1.** Students should title the graph "Favorite Kinds of Games"; list Card, Board, and Outdoor as the three categories; and draw bars—Card 6, Board 3, and Outdoor 8. **2.** 8 – 6 = 2, 2; **3.** 6 + 3 = 9, 9)

Model the Skill

- Hand out the Day 1 activity page. **Say:** *Graphs can be used to show data. What does the bar graph show?* (Favorite Ice Pop Flavors) Have students look at the labels. **Say:** *One axis tells us the different flavors and the other shows a scale of numbers. What are the flavors?* (Cherry, Grape, Lime, and Orange) **Say:** *To solve the first problem, find out how many like cherry. Slide your finger on the bar labeled Cherry and then down to the number scale. The bar is at what number?* (8) Repeat the process for grape. (6) **Say:** *To find how many students like cherry and grape, add the numbers.* Write the number sentence. **Ask:** *How many students like cherry and grape ice pops?* (14)

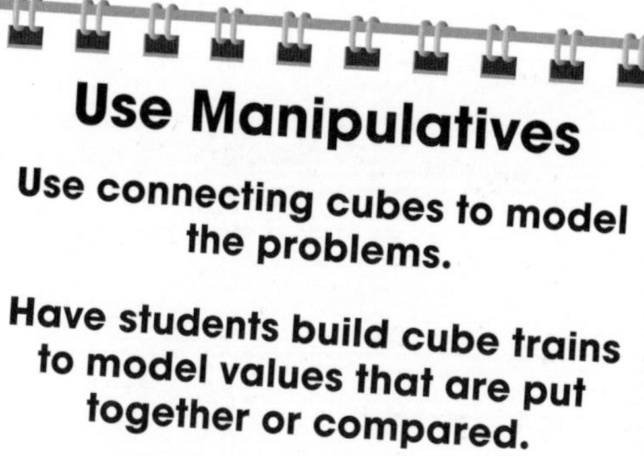

- Read aloud problem 2. **Ask:** *What number are you going to subtract from 22 to find how many students like another flavor better than lime?* (the number of students that like lime—2) Have students complete the number sentence to find the difference. (22 – 2 = 20)
- Now find the number of students that like orange and the number that like lime. Write a number sentence to compare the two amounts. (Answers: 6; 2; possible equation: 6 – 2 = 4, 4)

Make a Graph

Use the graph to solve each problem.

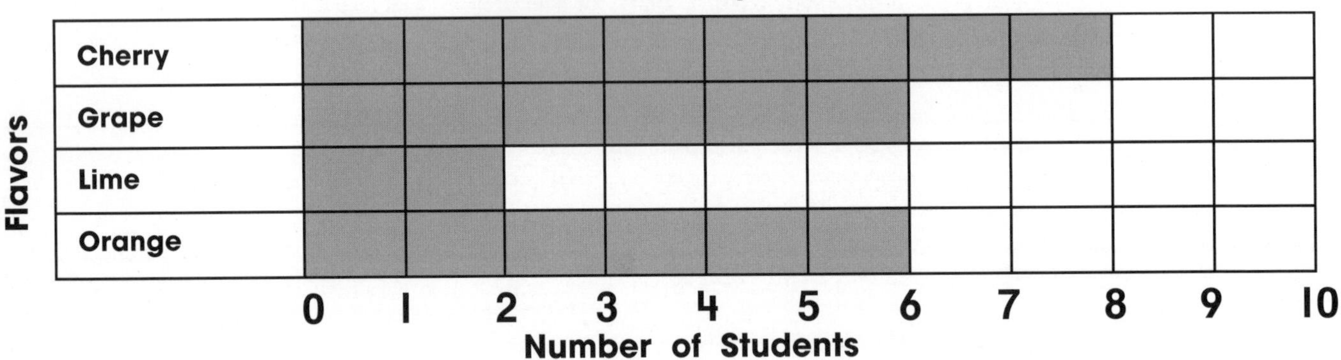

1. How many students like cherry and grape ice pops?

_______ students like cherry _______ students like grape

_______ + _______ = _______ students like cherry and grape

---

2. A total of 22 students are shown on the graph. How many students like another flavor better than lime?

22 − _______ = _______

_______ students like another flavor better than lime.

---

3. How many more students like orange than like lime?

_______ students like orange _______ students like lime

_______ ◯ _______ = _______

_______ more students like orange than lime.

☆ **Tell how you solved the problem.**

 Name ________________

Make a Graph

Use the graph to solve the problems.

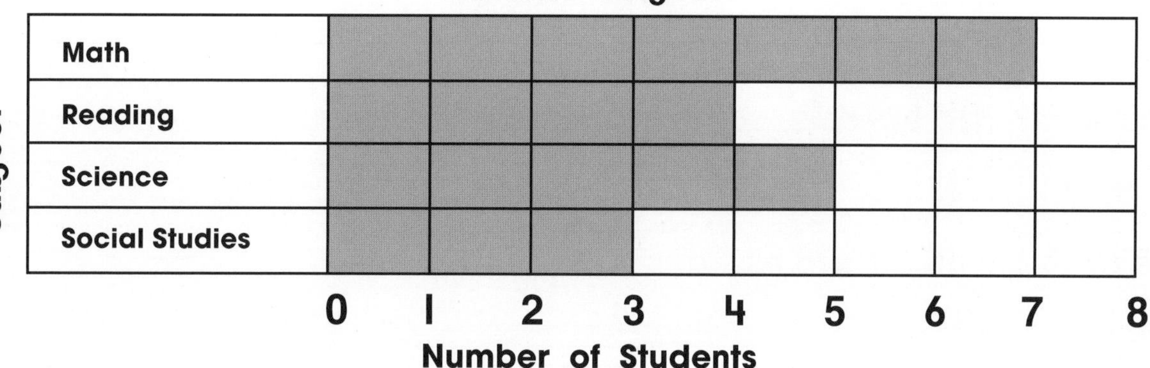

1. A total of 19 students are shown on the graph. How many students like another subject better than math?

_______ students like math 19 − _______ = _______

_______ students like another subject better than math.

---

2. How many students like reading and science?

_______ students like reading _______ students like science

_______ ◯ _______ = _______

_______ students like reading and science.

---

3. How many more students like math than like reading?

_______ ◯ _______ = _______

_______ more students like math than reading.

☆ **Tell how you find data on a bar graph.**

Name ______________________________

Make a Graph

Use the data to complete the bar graphs.

1

Favorite Sports	
Sport	**Number**
Soccer	𝍸 I
Baseball	II
Basketball	IIII
Football	𝍸 II

Favorite Sports

Sports										
Soccer										
Baseball										
Basketball										
Football										
	0 1	2	3	4	5	6	7	8	9	10

Number of Students

2

Pets We Have	
Pet	**Number**
Dog	𝍸 𝍸
Cat	𝍸 III
Fish	II

____

____

☆ **Tell how you know what to label each axis of the graph.**

 Name ______________________

Make a Graph

Use the data to complete the picture graphs.

Favorite Colors	
Color	**Number**
Red	IIII
Blue	~~IIII~~ II
Yellow	III
Green	~~IIII~~ I

Favorite Colors

Red	☺ ☺ ☺ ☺
Blue	
Yellow	
Green	

Key ☺ = 1 student

Favorite Snacks	
Snack	**Number**
Pretzels	5
Yogurt	3
Fruit	8

____

________	
________	
________	

Key ____________ = ____________

☆ **Tell how you know how many pictures to draw.**

Name ____________________

Assessment

Use the data to complete the graphs.

1

Favorite Kinds of Games	
Games	**Number**
Card	~~||||~~ |
Board	|||
Outdoor	~~||||~~ |||

____

Games								
Card								
Board								
Outdoor								
	0	1	2	3	4	5	6	7	8

Number of Students

2 How many more students like outdoor games than card games?

_____ ◯ _____ = _____

_____ more students like outdoor games than card games.

3 How many students like card games and board games?

_____ ◯ _____ = _____

_____ students like card games and board games.

☆ **Tell how you solved the problem.**

Overview Identify Shapes

Directions and Sample Answers for Activity Pages

Day 1	See "Model the Skill" below.
Day 2	Read the directions aloud. Remind students that some shape names indicate the number of sides and angles. Encourage students to match the names to the shapes that they know before matching the names to the shapes they are not as sure about. (Answers: Students should draw a line to: **1.** triangle; **2.** quadrilateral; **3.** pentagon; **4.** hexagon; **5.** cube.)
Day 3	Read the directions aloud. Review the key with students and have them make a crayon mark next to each shape name. Encourage students to choose one shape/color and color all the like shapes, such as the triangles, rather than coloring each shape in order. (Students should color 3 quadrilaterals yellow, 3 triangles blue, 3 cubes green, and 2 pentagons red.)
Day 4	Read the directions aloud. Point out that the shape names are given in a word box. Remind students that the named shapes are two-dimensional or flat, and that they all have straight sides. (Answers: Check student drawings. **1.** quadrilateral; **2.** hexagon; **3.** triangle; **4.** pentagon)
Day 5	Read the directions aloud. Observe as students complete the page. Do students identify the correct shape? Do they connect the shape name to the number of sides or angles? Use your observations to plan further instruction and review. (Answers: **1.** hexagon; **2.** quadrilateral; **3.** cube; **4.** Check students' drawings.)

Model the Skill

- Hand out the Day 1 activity page and red and blue crayons. Display models of triangles, quadrilaterals, pentagons, hexagons, and cubes. **Ask:** *Which one of these shapes is most different from the other shapes?* (a cube) Hold up a cube and explain that it is a three-dimensional shape while the other shapes are two-dimensional, or flat, shapes. **Ask:** *How many flat sides, or faces, does a cube have?* Have two volunteers work together to determine the number of faces. (6) **Ask:** *What shape are the faces of a cube?* (square)

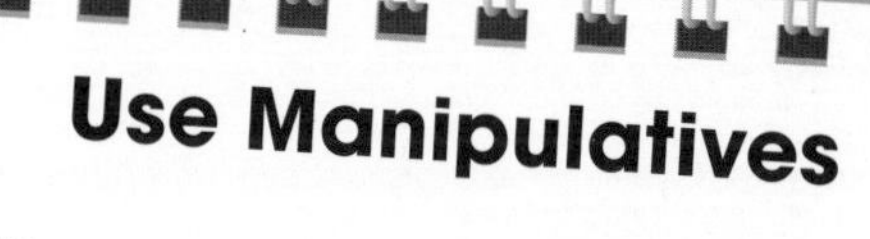

Use Manipulatives

Use pattern blocks, tangram pieces, and other shapes.

Have students count the number of sides and angles.

Guide students to link those numbers to the shape names.

- **Say:** *We are going to count the number of sides and angles of the four shapes shown. What is the shape shown in problem 1?* (triangle) **Say:** *Use a blue crayon to trace the sides of the triangle. How many sides did you trace?* (3) Allow tactile learners to touch the triangle models. **Say:** *To count the angles, use a red crayon to make a mark on each angle. How many angles are there?* (3) Explain to students that *tri-* means three, and a triangle is a shape with three angles.
- Have students trace and mark the sides (4) and mark the angles (4) in problem 2. **Say:** *Think about other words that you know that begin with quad.* Quad *means four. A quadrilateral has four sides and four angles.* Point out various quadrilaterals including a square, a rectangle, a rhombus, a trapezoid, and other parallelograms.
- Help students complete the activity page by tracing the sides of each shape and then marking the angles of each shape. Explain that *penta-* means five and *hexa-* means six. Guide students to link the names to the number of sides and angles of each shape. (Answers: **3.** 5, 5; **4.** 6, 6)

 Name ________________________

Identify Shapes

List the number of sides for each shape. Then list the number of angles.

1 triangle

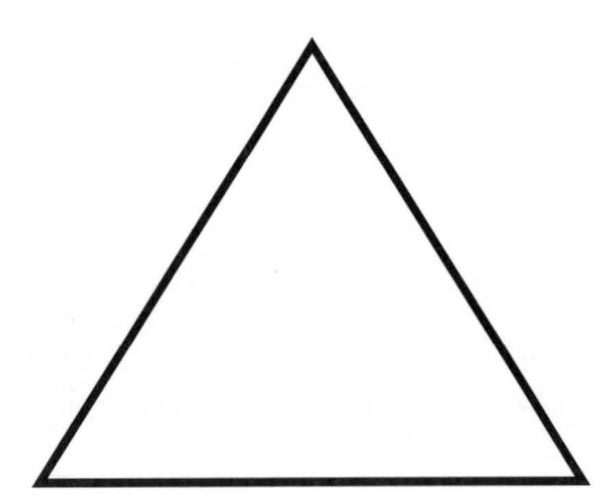

sides ________

angles ________

2 quadrilateral

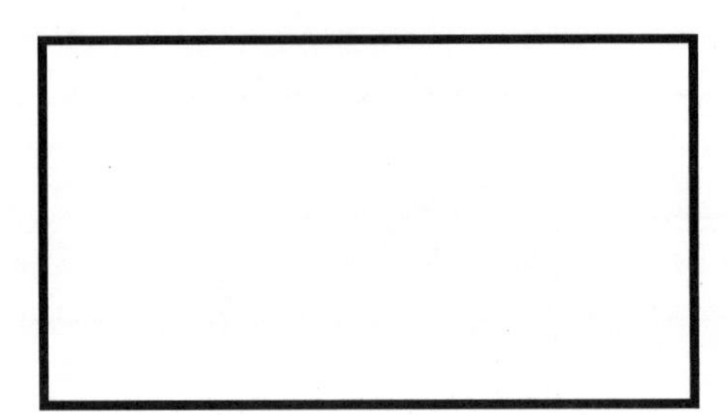

sides ________

angles ________

3 pentagon

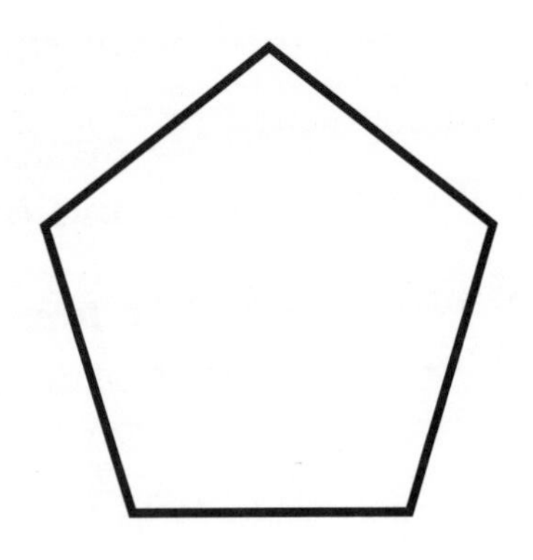

sides ________

angles ________

4 hexagon

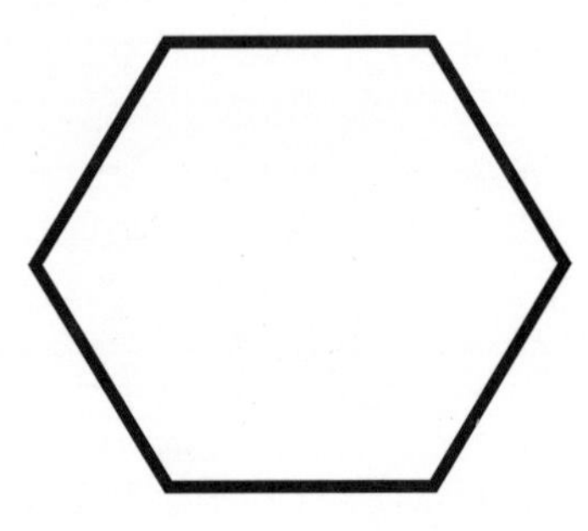

sides ________

angles ________

☆ **Tell how the number of sides matches the shape name.**

Name ____________________

Identify Shapes

Match each shape to its name.

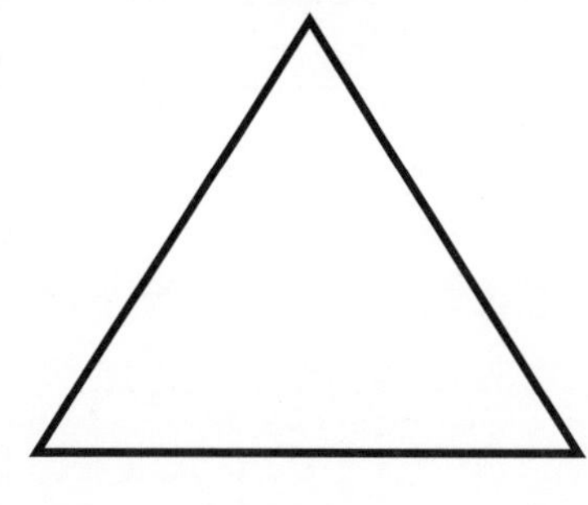

pentagon

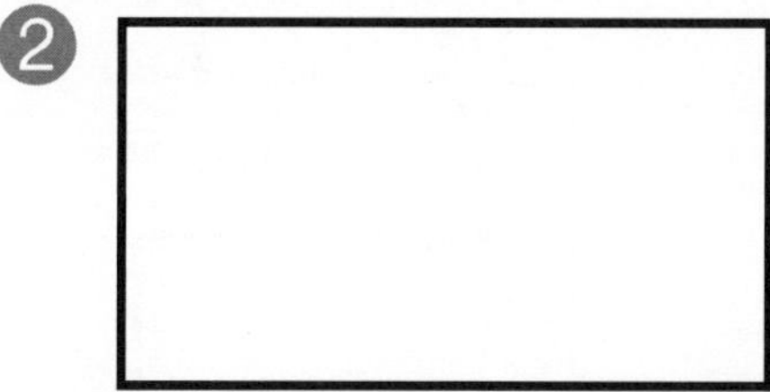

triangle

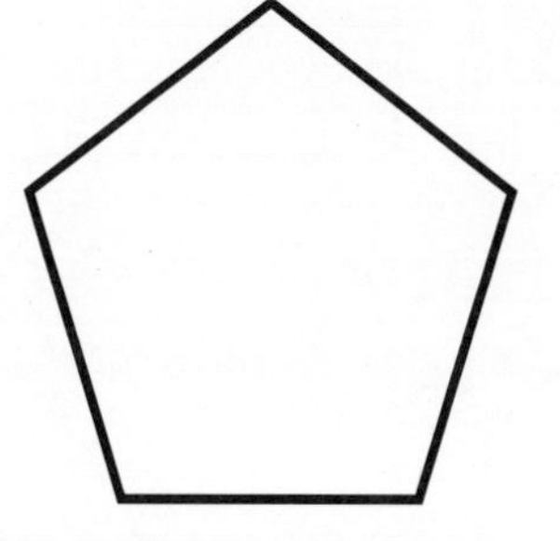

quadrilateral

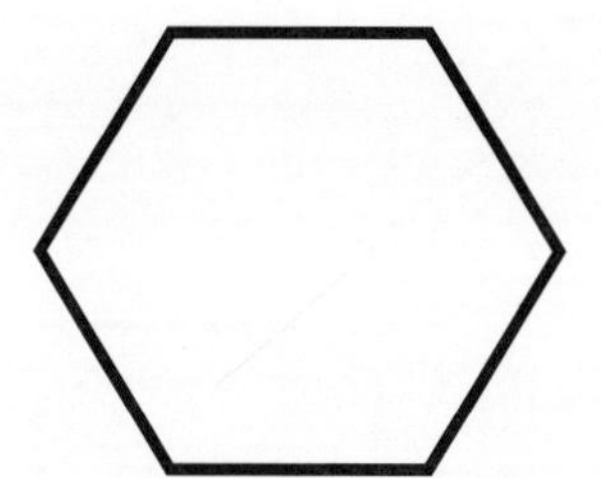

cube

5
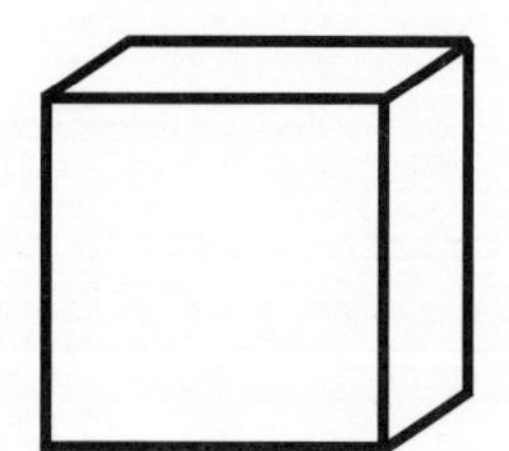

hexagon

 **Tell how you know the name of a shape.**

Name ______________________________

Identify Shapes

Color the shapes according to the key.

triangles	=	blue
cubes	=	green
pentagons	=	red
quadrilaterals	=	yellow

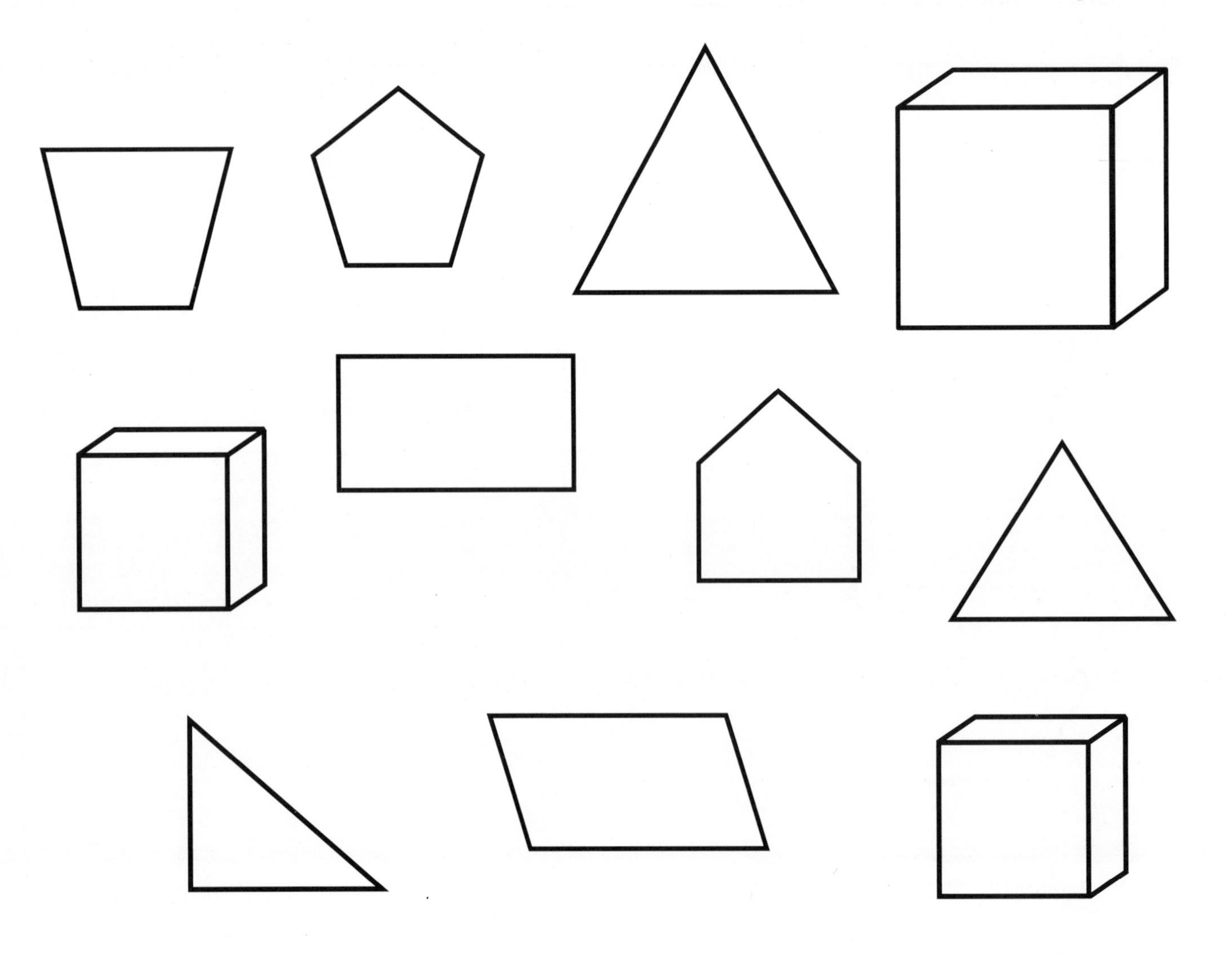

☆ **Tell how you know which shapes are quadrilaterals.**

Name ____________________

Identify Shapes

Draw each shape described. Write its name.

triangle
quadrilateral
pentagon
hexagon

1. I have four sides and four angles. What am I?

 ____

2. I have six sides and six angles. What am I?

 ____

3. I have three sides and three angles. What am I?

 ____

4. I have five sides and five angles. What am I?

 ____

☆ **Tell how you know which shape is described.**

 Name ______________________

Assessment

Match each shape to its description.

1. **Shape with 6 angles.** **quadrilateral**

2. **Shape with 4 sides.** **cube**

3. **Shape with 6 faces.** **hexagon**

Draw the shape on the line.

4. pentagon ______________________

☆ **Tell how you solved the problem.**

Overview Parts of Shapes

Directions and Sample Answers for Activity Pages

Day 1	See "Model the Skill" below.
Day 2	Read the directions aloud. Remind students that equal shares are the same size. Point out that two shapes in each problem show equal shares while one shape does not. Students should note that equal shares of identical rectangles can be different shapes. (Answers: **1.** Students should circle the first and third shapes. **2.** Students should circle the second and third shapes. **3.** Students should circle the first and second shapes. **4.** Students should circle the first and third shapes.)
Day 3	Read the directions aloud. Point out that there is a rectangle and a circle that matches each description. Remind students that the parts must be equal in size. Encourage students to look for number words within each description to help match shapes. (Answers: **1.** third rectangle, second circle; **2.** first rectangle, first circle; **3.** second rectangle, fourth circle)
Day 4	Read the directions aloud. Point out that the shapes are divided into equal shares. Students should color shares of each shape to match each description. Guide students to see that two-fourths are the same as one-half. (Answers: **1.** Students should color one part. **2.** Students should color one part. **3.** Students should color one part. **4.** Students should color two parts.)
Day 5	Read the directions aloud. Observe as students complete the page. Do students draw the correct number of equal shares? Do they understand that halves show two parts? Do they see that four-fourths is one whole? Use your observations to plan further instruction and review. (Answers: **1.** Check drawings. **2.** Check drawings. **3.** Students should circle the second circle. **4.** Students should circle the third rectangle.)

Model the Skill

- Hand out the Day 1 activity page and a rectangular sheet of paper.
- **Say:** *Imagine that you and a friend both want to draw, but you have only one sheet of paper. What can you do so that you each have an equal share of the paper?* (Answers will vary. Possible answer: Fold the paper in half and cut along a fold.) Guide students to fold one sheet of paper in half, matching corners. **Say:** *Each share of the paper is the same size. The shares are equal. How is this paper divided?* (into halves) Have students draw a line on the rectangle in problem 1 to show two equal shares. **Say:** *Each half of the rectangle is one-half. Two-halves are the same as one whole.*
- Have students look at problem 2. **Say:** *We can fold the paper again to make four equal shares.* Guide students to match corners, fold, and then open the paper to see the four sections. **Say:** *The paper shows fourths. Four-fourths are the same as one whole.* Guide students to draw lines on the rectangle in problem 2 to show four equal shares. Point out the word **four** in fourths.
- Help students complete the activity page by drawing lines to show equal shares. If necessary, have students fold another sheet of paper to see the equal shares. (Check students' drawings.) Point out that the shape in problem 4 shows thirds and that three-thirds is the same as one whole.

Use Shapes

Use paper circles and rectangles.

Fold them to show equal shares. Count the equal shares.

Color shares to show fractional parts—one-third, one-fourth, etc.

Name ______________________

Parts of Shapes

Draw lines to show equal shares.

1. Show two equal shares.

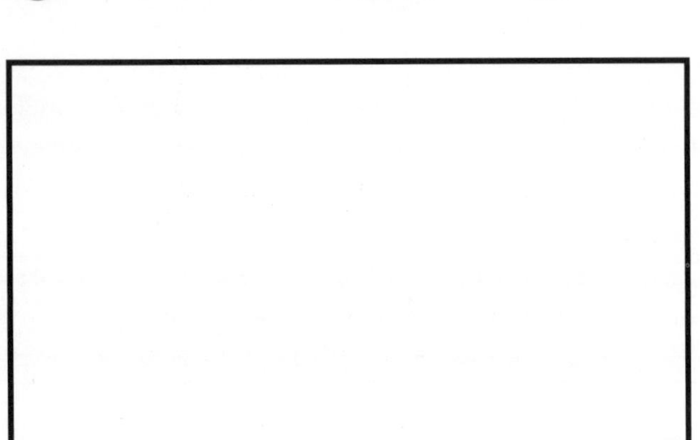

2. Show four equal shares.

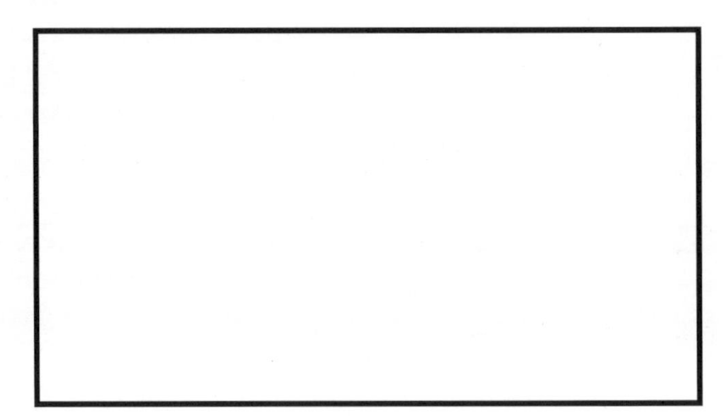

3. Show two equal shares.

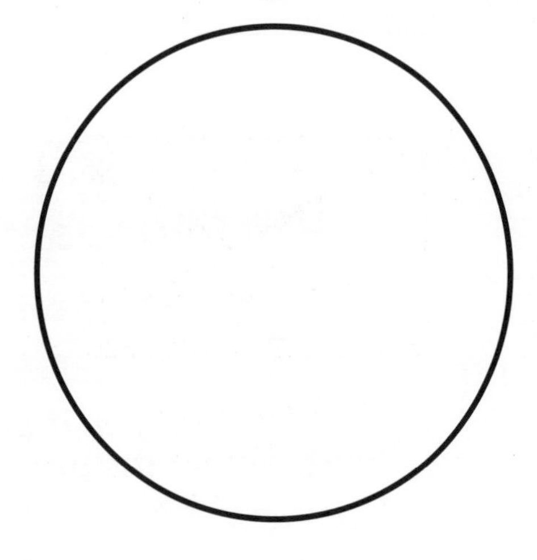

4. Show three equal shares.

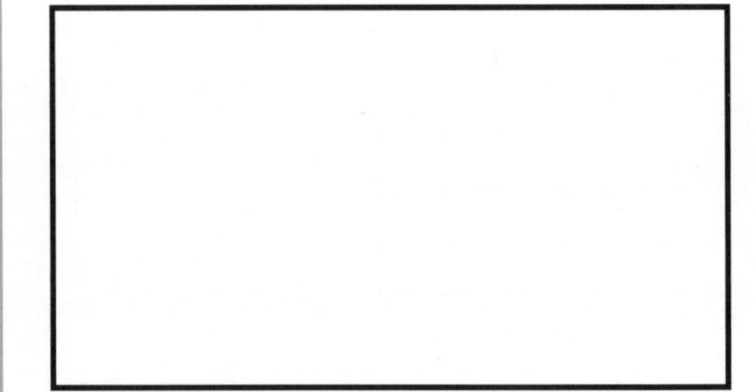

☆ **Tell how you know where to draw the lines.**

Name ____________________

Parts of Shapes

Circle the shapes that show equal shares.

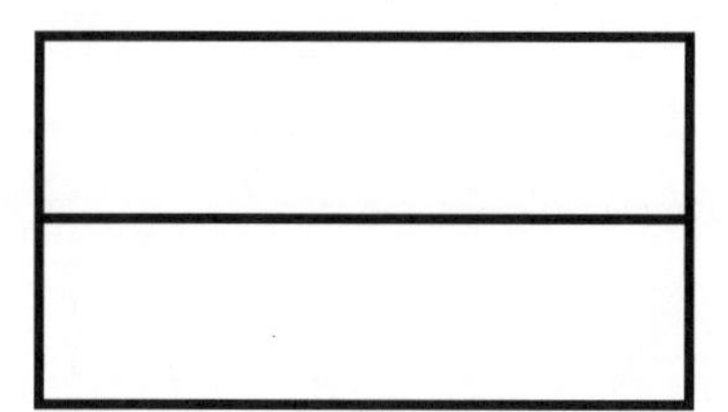 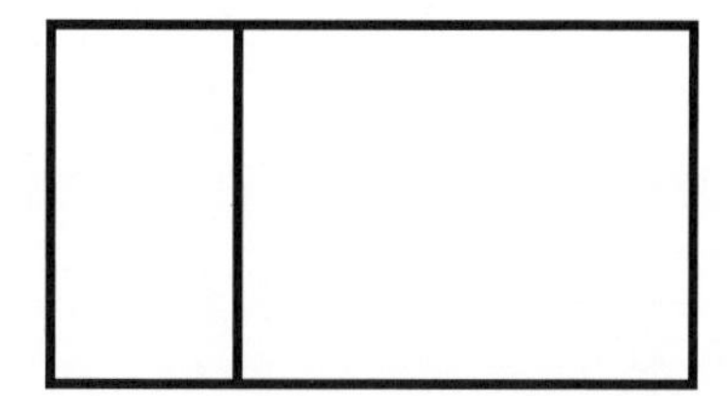 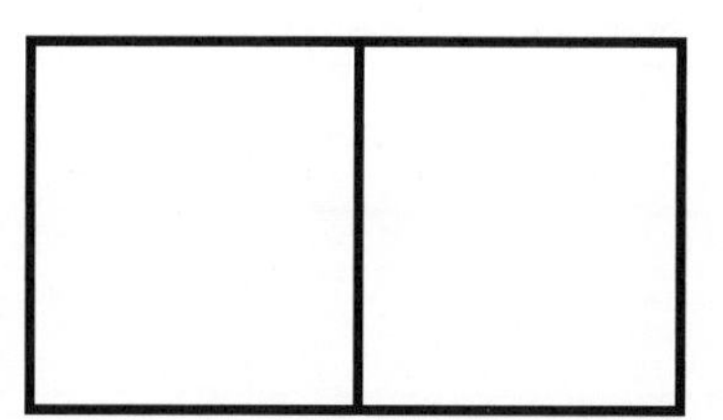

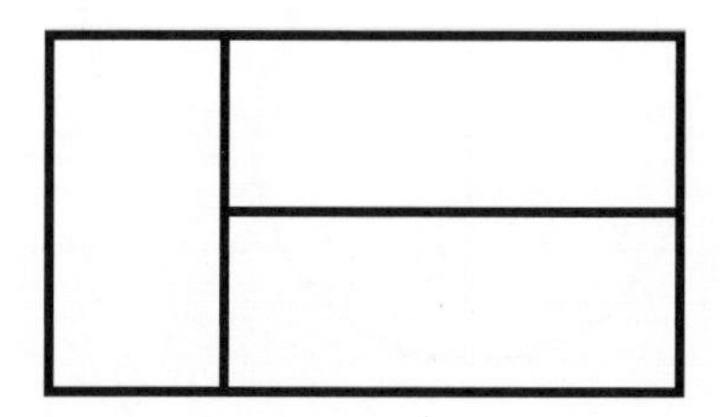 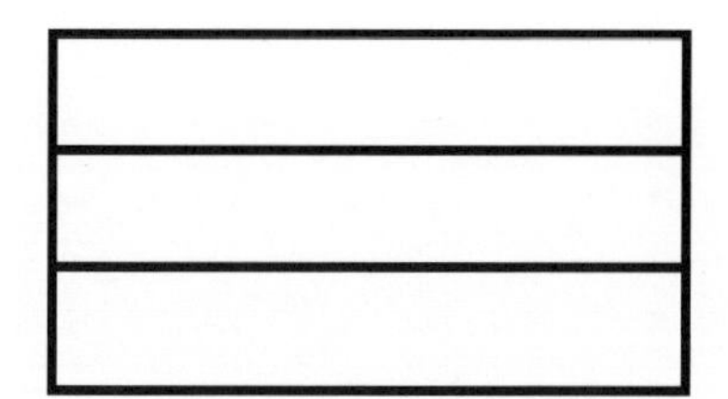 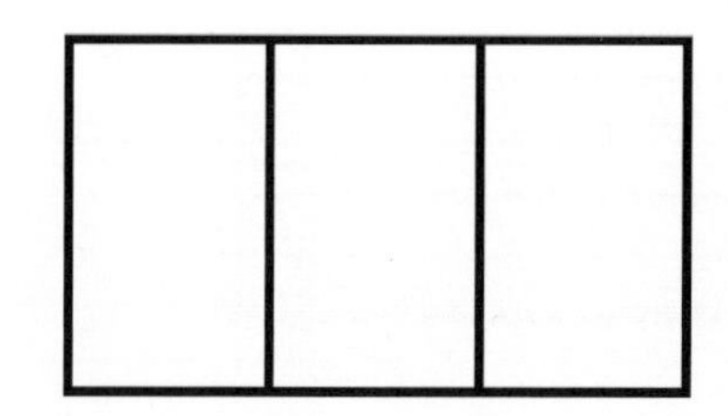

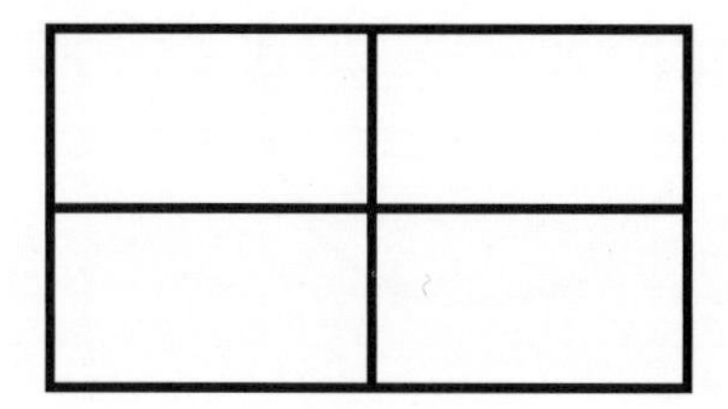 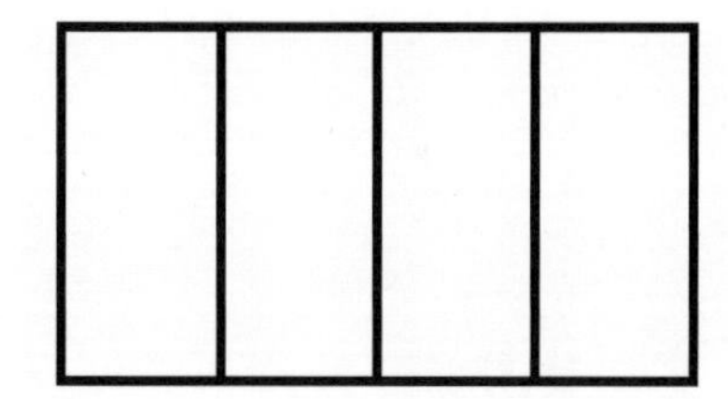 

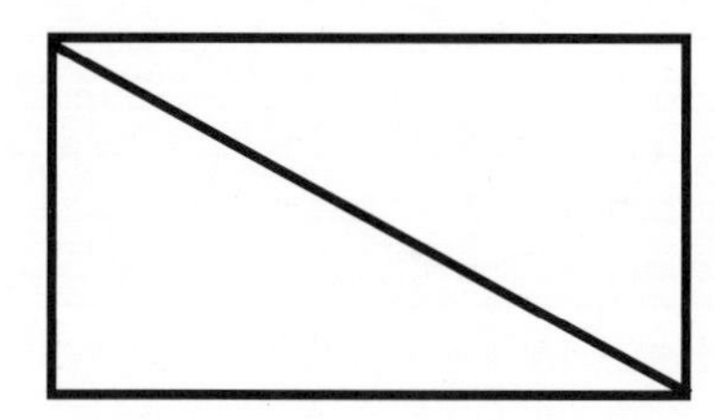 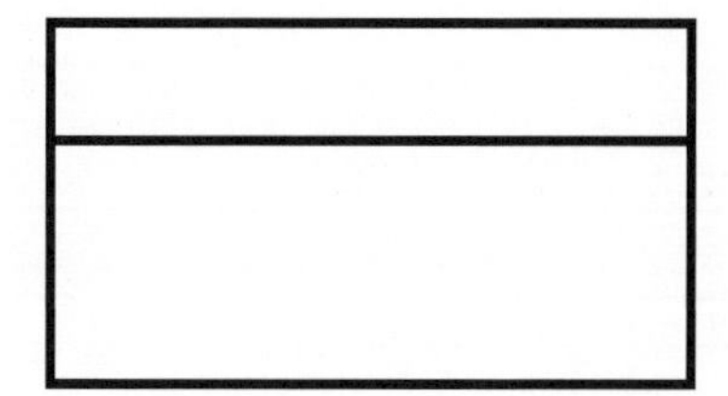 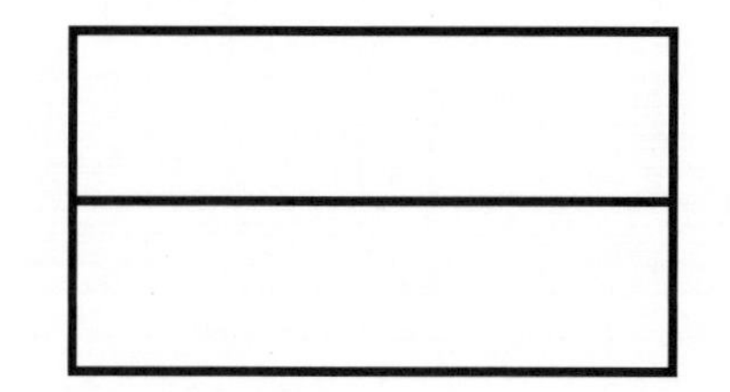

 **Tell how equal shares of the same shape can be different shapes.**

Parts of Shapes

Draw lines to match each shape to its description.

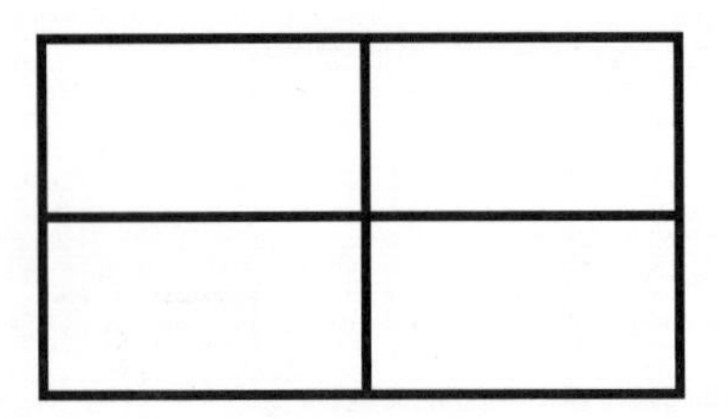

halves

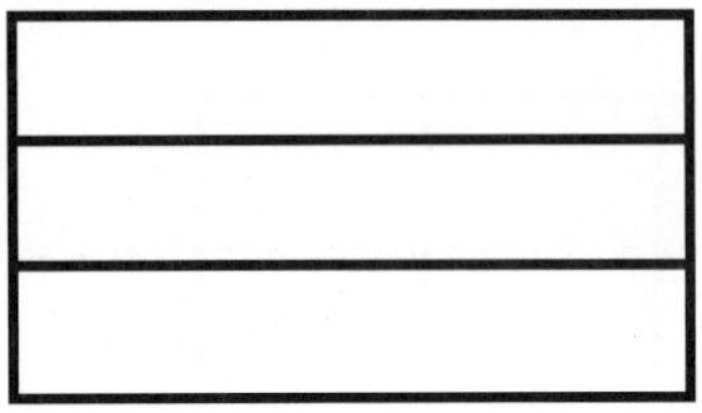

fourths

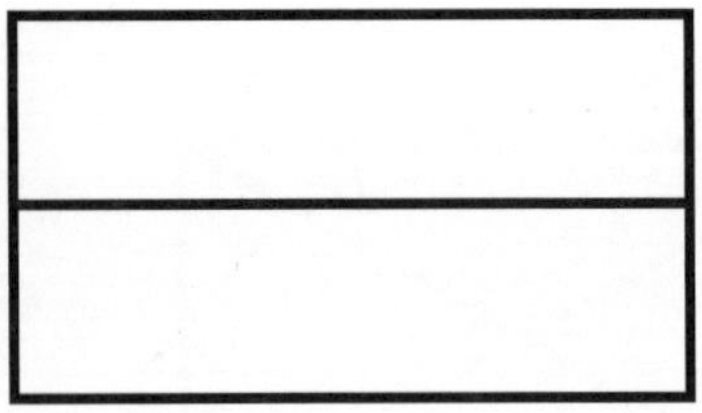

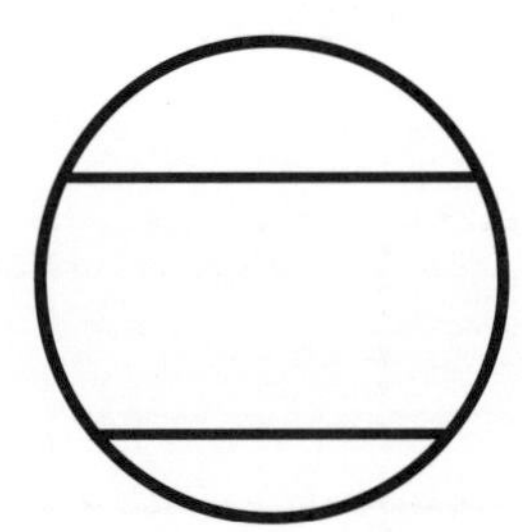

thirds

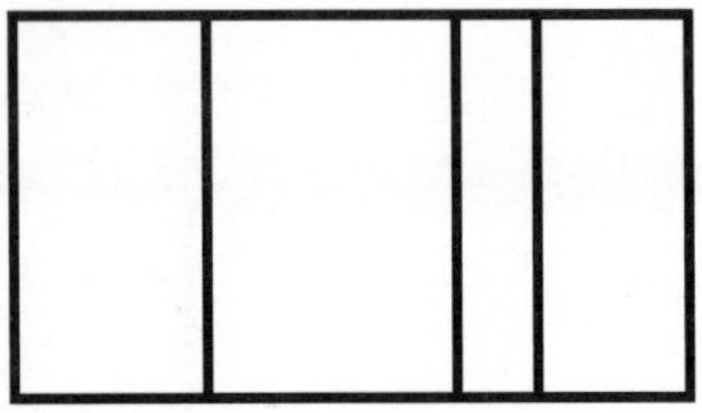

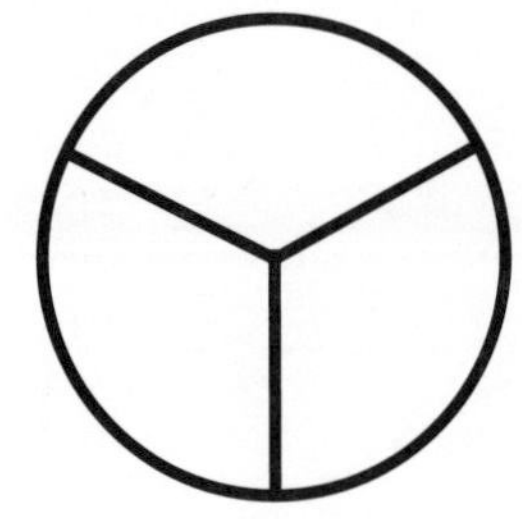

☆ **Tell how you know which description matches each shape.**

Name ___________________________

Parts of Shapes

Color to show the shares.

1 one-half

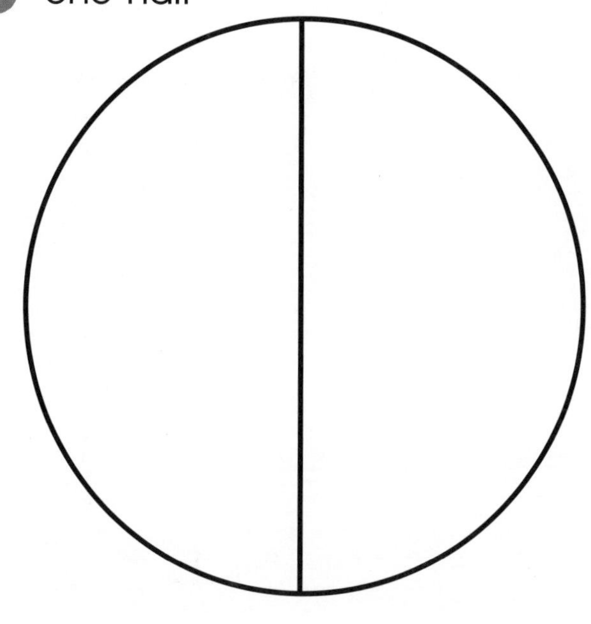

2 one-fourth

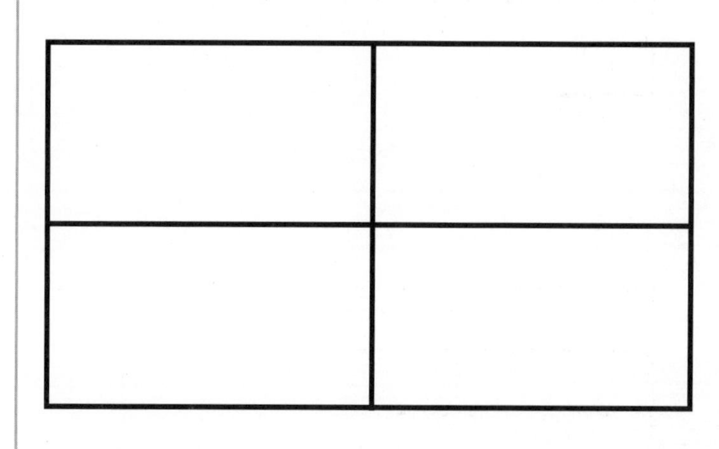

3 one-third

4 two-fourths

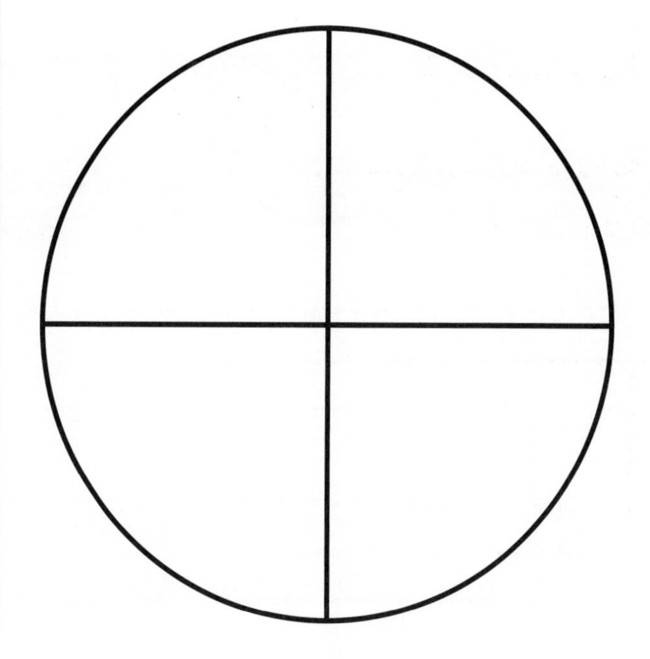

☆ **Tell how two-fourths is like one-half.**

Name ______________________________

Assessment

Solve each problem.

1 Draw lines to show four equal shares.

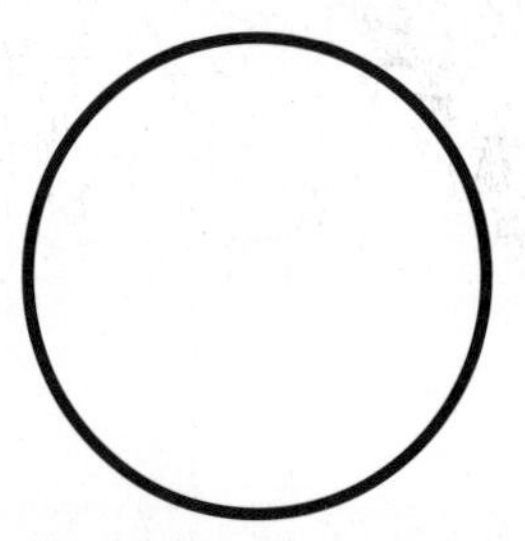

2 Draw lines to show three equal shares.

3 Circle the shape that shows halves.

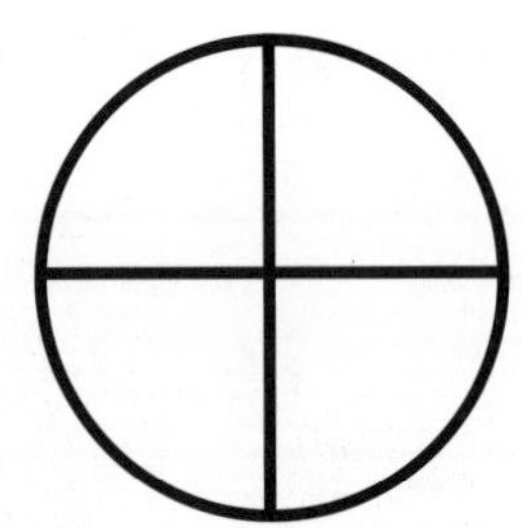

4 Circle the shape that shows four-fourths.

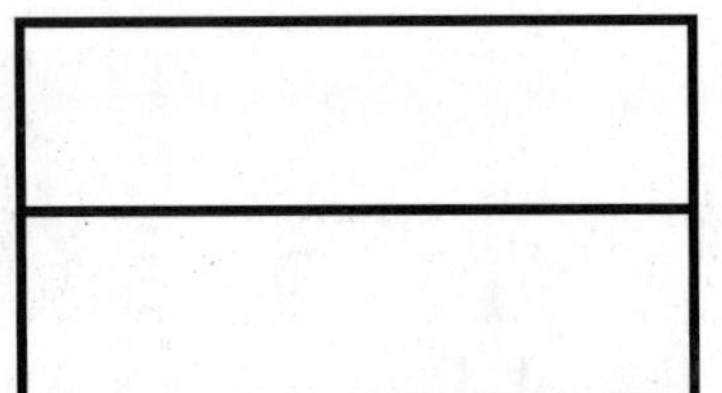
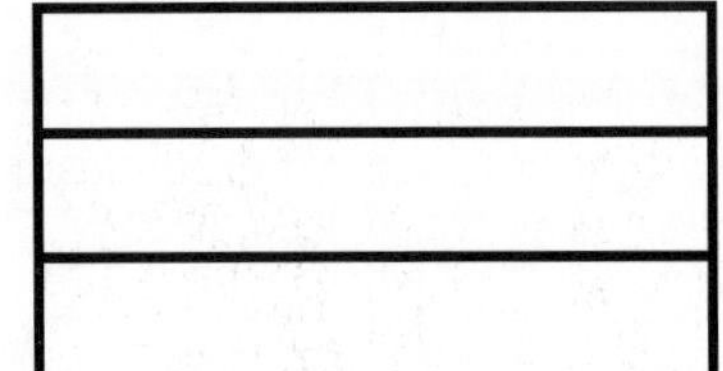
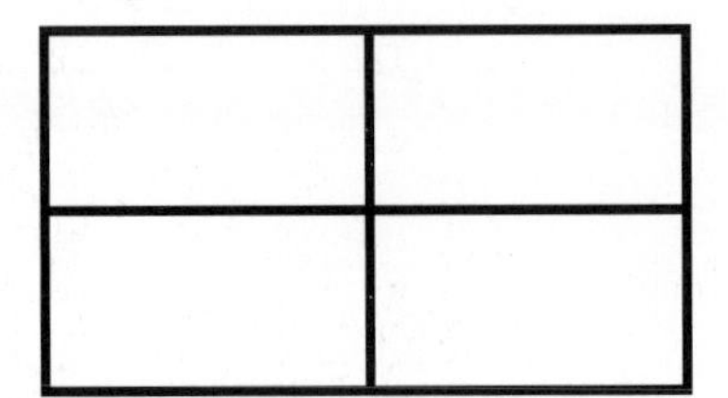

☆ **Tell how you solved the problem.**

